Y0-BYV-093

A Guide to U.S. Government
Scientific and Technical Resources

A GUIDE TO U.S. GOVERNMENT SCIENTIFIC AND TECHNICAL RESOURCES

RAO ALURI
Assistant Professor
Division of Library and
Information Management
Emory University
Atlanta, Georgia

JUDITH SCHIEK ROBINSON
Associate Professor
School of Information and
Library Studies
State University of New York
at Buffalo
Buffalo, New York

LIBRARIES UNLIMITED, INC.
Littleton, Colorado
1983

LIBRARIES UNLIMITED, INC.
P.O. Box 263
Littleton, Colorado 80160-0263

Library of Congress Cataloging in Publication Data

Aluri, Rao.
A guide to U.S. government scientific and technical resources.

Includes indexes.
1. Science--Information services--United States--Directories. 2. Technology--Information services--United States--Directories. I. Robinson, Judith Schiek, 1947- . II. Title. III. Title: Guide to US government scientific and technical resources.
Q224.3.U6A43 1983 506'.073 83-14991
ISBN 0-87287-377-3

Libraries Unlimited books are bound with Type II nonwoven material that meets and exceeds National Association of State Textbook Administrators' Type II nonwoven material specifications Class A through E.

To our spouses
Mary Reichel
Bruce Robinson

ACKNOWLEDGMENT

The authors take this opportunity to thank the Council on Library Resources for providing financial support for the preparation of this guide in the form of fellowships to the authors. Also, the authors benefited from the critical comments made by Professor Esther M. Stokes, Affiliated Assistant Professor, Emory University, Division of Library and Information Management, and Mr. LeRoy C. Schwarzkopf. Their criticisms, we believe, improved the usefulness of this guide. We extend our sincere thanks to both of them for their time and expertise. We express our gratitude to Professor K. Subramanyam for contributing the chapter on standards and specifications, to Professor Hilda Feinberg for preparing the index, to Ms. Margaret V. Norman for proofreading, and to Ms. Joanna Goldman for inputting the text into a word processor.

TABLE OF CONTENTS

LIST OF ILLUSTRATIONS

1

INTRODUCTION

THE GROWING NEED FOR A GUIDE TO FEDERALLY SPONSORED RESEARCH

The United States government is one of the largest producers of scientific and technical information in the world. This information is generated by a vast network of scientific organizations, maintained or supported by the federal government, which engage in scientific research and development (R&D) activities. The organizations that perform research with federal support include federal government laboratories, universities and colleges, industrial firms, nonprofit organizations, and federally funded research and development centers.[1]

National Institutes of Health, which is a part of the Department of Health and Human Services, for example, consists of fourteen institutes and research centers engaged in medical research. Examples of these institutes are the National Cancer Institute, National Eye Institute, and National Institute of Mental Health.

The significance of federal intramural research is illustrated by the fact that four of the Nobel Laureates are NIH scientists: Marshall W. Nirenberg (1968);

Julius Axelrod (1970); Christian B. Anfinsen (1972); and D. Carleton Gajdusek (1976). Anfinsen received his prize for his work in chemistry; the others received their prizes for work in physiology or medicine.

Research publication output of NIH scientists is evidenced from the publication *Scientific Directory 1982: Annual Bibliography 1981.** Hundreds of colleges and universities such as Massachusetts Institute of Technology, Johns Hopkins University, Columbia University, and Stanford University are actively engaged in research and derive a significant amount of support for this activity from the federal government.

Nearly half of federal obligations for research and development is spent in commercial industrial organizations. Much of the federally supported R&D activity in the industrial firms is oriented to national security. Industries that receive the greatest share of federal monies include aircraft and missiles, electrical equipment, and communication.

Many nonprofit organizations also derive a significant part of support for their research activities from the U.S. government. Examples of such nonprofit organizations are Battelle Memorial Institute and the Rand Corporation. There are about thirty-five Federally Funded Research and Development Centers (FFRDCs) which are typically administered by colleges and universities, industrial firms, and other nonprofit organizations.[2]

The FFRDCs are R&D-performing organizations that perform "actual research and development or R&D management either upon direct request of the federal government or under a broad charter from the federal government, but in either case under the direct monitorship of the federal government."[3] Examples of such organizations are E.O. Lawrence Berkeley Laboratory (administered by the University of California), Oak Ridge National Laboratory (administered by Union Carbide Corp.), and Solar Energy Research Institute (a nonprofit organization).

This network of research organizations is expected to spend $77.6 billion in 1982 for research and development. Nearly half of this support will have been received from the federal government.[4] Again, the pervasive influence of the quality of federal research support is evidenced by the fact that at least seventy Nobel Laureates such as Hamilton O. Smith (1978), David Baltimore (1975), Herbert G. Brown (1979), and Otto H. Warburg (1931) were recipients of support from federal agencies such as the National Cancer Institute.

DISTRIBUTION OF FEDERAL RESEARCH INFORMATION

This extensive federal support for scientific research and development results in a great amount of scientific and technical information, which is the focus of this volume. The federal government, for example, is estimated to have spent $529.9 million for scientific and technical information activities in 1978.[5] These information dissemination activities include publication and distribution ($161.1 million), documentation, reference, and information services ($262.9 million),

*U.S. Department of Health and Human Services, National Institutes of Health, *Scientific Directory 1982: Annual Bibliography 1981* (NIH Pub. No. 82-4) (Washington, DC: Government Printing Office, 1982), 467p. (HE20.3017:982).

symposia and audiovisual media ($31.6 million), and research and development in information sciences* ($74.3 million).[6]

The publication and distribution activities include the publication of primary literature such as journal articles, "technical reports, patents, dissertations, data compilations, proceedings of conferences and symposia, specifications and manuals used in the R&D process, monographs, [and] serials."[7] These activities also include the publication and distribution of secondary and tertiary publications such as "abstracts, indexes, dictionaries, textbooks, handbooks, bibliographies, reviews, encyclopedias, [and] directories."[8]

Nearly 6 percent of the publications and distribution activities went to "support of publications" category which includes "all page charges paid out of federal funds to primary journals," and other "support mechanisms to assure the viability of certain publications."[9] The services of libraries, specialized information centers and other such organizations account for over 90 percent of expenditures in the category of documentation, reference, and information services.[10]

The federal government also supports travel and participation in scientific meetings, production and distribution of videotapes, motion pictures and other audiovisual materials as well as research and development activities in library and information sciences. Nearly 70 percent of these expenditures for scientific and technical communications were accounted by three federal agencies; namely, the departments of Defense, Commerce, and Health, Education and Welfare.[11]

COMPREHENSIVE GUIDE NEEDED

Unfortunately, this extensive scientific information activity of the U.S. government has not attracted adequate attention from the library profession. No systematic effort has been made to show the breadth and diversity of this vast amount of scientific and technical information produced by the federal government. Currently, two types of publications endeavor to fill this need: guides to the literature of science and technology subjects, and guides to federal government documents.

The guides to the literature of science and technology typically explain the structure of the literature of the given subject in terms of primary, secondary, and tertiary publications, and usually include separate chapters on government documents.[12] However, this segregation of government publications as a format distinct from other formats such as journals fails to highlight that a vast number

*A recent King Research study presents radically different figures. The report states that "the Federal government expends an estimated $6.4 billion on scientific and technical information transfer. These expenditures include $175 million for government-published journals and other periodicals, books, technical reports and other materials (such as patent documents); $1.2 billion for library and abstracting and indexing activities; $1 billion for numeric data bases; $1.2 billion for authorship expenses including author salaries paid by the government or under contract or grant; $2.3 billion for reading by scientists paid by the government through salaries or under contract or grant; and $502 million for other scientific and technical information activities." This is in addition to an annual budget of $490 million for a network of 196 federally funded information organizations which participated in the King Research survey. See: Donald W. King and Dennis D. McDonald, *Federal and Non-Federal Relationships in Providing Scientific & Technical Information: Policies, Arrangements, Flow of Funds and User Charges* (Rockville, MD: King Research, Inc., March 1980), p. 7.

of primary, secondary, and tertiary scientific publications are issued by the federal government or funded by federal monies.

There is an equally large number of guides and bibliographies that concentrate upon federal publications.[13] These guides, however, tend to emphasize the governmental structure, while playing down the subject aspects of the publications mentioned. Even when a subject approach is used, the emphasis is markedly away from scientific and technical publications. Bibliographies of federal publications or reference books, of course, are not designed to serve as guides to the federal scientific and technical literature.

In other words, the existing guides to literature fail to emphasize the intricate relationship between the federal publications and the literature of science and technology. The federal government, for example, is the predominant source for certain types of scientific literature such as patents and technical reports. Consequently, there is a need for a guide to federal scientific and technical literature that bridges the bibliographic gap between guides to federal publications and the guides to scientific and technical literature. This publication, *A Guide to U.S. Government Scientific and Technical Resources*, represents an attempt to fill that need.

The guide is designed to direct the reader to the vast and diverse scientific and technical information available from the United States government. It is hoped that the guide will be useful to science and engineering information personnel and to the students of government documents and/or science literature courses in library schools.

HOW TO USE THIS GUIDE

This guide is organized to reflect the normal patterns of scientific communications where the information goes through a number of successive stages.[14] These stages include seeking funds for research, announcement of information through primary media such as technical reports, journal articles, patents, and audiovisual media. These stages are followed by the repackaging of the primary literature for retrieval; namely, indexes and abstracts, and machine-readable data bases. Consequently, chapters in this guide are arranged so that the reader encounters various forms of primary literature first, and then secondary literature. Although information analysis centers are not a form of scientific literature, a chapter is devoted to them because of the unique functions they serve in terms of collection, analysis, and evaluation of scientific and technical information.

Each chapter focuses upon a category of information; for example, patents and trademarks, and discusses its place in the flow of scientific and technical communication and mentions major sources of information in that category. If non-government sources are available, which provide access to government-produced information in this category, those sources are also described. Similarly, non-bibliographic sources are listed where appropriate. The objective of this guide is to provide a strategy for locating both current and future scientific and technical information that use federal government sources.

In this connection, one point needs to be emphasized: it is almost certain that by the time this guide is published, many specific sources mentioned will have gone through several changes. Some of the publications may have been discontinued or may have been assigned broadened or lessened responsibilities. With every change of administration at the federal level, there will be a number of such discontinuities. Consequently, the reader is advised that the present publication is a guide to the literature and is *not* a current bibliography. The reader must consult current sources of information such as the *Monthly Catalog of U.S. Government Publications* and *Government Reports Announcements & Index* to ascertain the most recent publications in his/her areas of interest.

In any event, in spite of changes that are likely to take place at a specific level such as the discontinuing of some publications and disbanding of specific agencies, the structure presented in this guide will still be useful. That is, the federal government, for the foreseeable future should be heavily involved in national scientific research and development activities.[15] It will continue to generate a vast amount of scientific information. Looking at that information from the perspective of primary and secondary forms of scientific literature will make it easier to comprehend the extent of federally generated scientific information.

REFERENCES

1. National Science Foundation, *Federal Funds for Research and Development: Fiscal Years 1980, 1981, and 1982*, Volume XXX (NSF 81-325) (Washington, DC: 1981), pp. 2-3.
2. Ibid., pp. 8-9.
3. Ibid., p. 3.
4. Jules J. Duga and W. Halder Fisher, *Probable Levels of R & D Expenditures in 1982: Forecast and Analysis* (Columbus, OH: Battelle Memorial Institute, December 1981), p. 5.
5. National Science Foundation, *Federal Funds for Research, Development, and Other Scientific Activities: Fiscal Years 1976, 1977, and 1978*, Volume XXVI (NSF 78-300) (Washington, DC: Government Printing Office, 1977), p. 42.
6. Ibid., p. 43.
7. Ibid., p. 51.
8. Ibid.
9. Ibid.
10. Ibid., p. 43.
11. Ibid., p. 42.
12. Examples of such guides include: E. J. Crane et al., *A Guide to the Literature of Chemistry*, 2d ed. (New York: Wiley, 1957); D. N. Wood, *Use of Earth Sciences Literature* (Hamden, CT: Shoe String, 1973); and Saul Herner, *A Brief Guide to Sources of Scientific and Technical Information*, 2d ed. (Washington, DC: Information Resources Press, 1980).

13. Joe Morehead, *Introduction to United States Public Documents*, 3d ed. (Littleton, CO: Libraries Unlimited, 1983) and Laurence F. Schmeckebier and Roy B. Eastin, *Government Publications and Their Use*, 2d rev. ed. (Washington, DC: Brookings Institution, 1969).

14. William D. Garvey and Belver C. Griffith, "Communications and Information Processing within Scientific Disciplines: Empirical Findings for Psychology," in William D. Garvey, *Communication: The Essence of Science* (Oxford: Pergamon Press, 1979), pp. 127-47.

15. U.S. National Science Foundation, *1990 R&D Funding Projections* (NSF 82-315) (Washington, DC: July 1982).

2

GRANTS, AWARDS, FELLOWSHIPS, AND SCHOLARSHIPS

The initial step of a research and development endeavor is for a scientist or an engineer to spend a year or more developing the research idea, writing the proposal, and submitting it to one or more of the federal agencies. This is especially true if the scientist works for non-federal agencies such as colleges and universities and is seeking support from agencies such as the National Science Foundation and National Institutes of Health.

Generally, during much of this phase, the information generated by the scientist may not be accessible to fellow scientists except through informal channels such as personal contacts. Such an unavailability, clearly, leaves a significant gap in the scientific communication process. For example, if a scientist is unaware of the fact that a fellow scientist has already obtained support from a federal agency to work on a certain research topic, then he or she may duplicate the work of the fellow scientist and thereby waste his/her resources. On the other hand, the awareness of such information will be valuable in terms of developing informal contacts and research consultation among the scientists of similar interests.

AVAILABILITY OF FEDERAL GRANT INFORMATION

In this respect, various agencies of the federal government issue several publications of use to the scientist. These publications help the scientist in identifying the agencies likely to fund his/her research and their grant-awarding policies and procedures, and in identifying the individuals and institutions active in his/her area of research.

The federal government releases information on grants, awards, fellowships, and scholarships at numerous levels. Many agencies regularly issue pamphlets and fliers describing their own grant programs. Some of these can be identified by searching the *Monthly Catalog of United States Government Publications* (Washington, DC: Government Printing Office). Similarly, by checking the *U.S. Government Manual* (Washington, DC: Government Printing Office, Annual), a researcher can identify agencies and subagencies that monitor and announce grant opportunities in various subject areas and can request them to send recent announcements. By writing individual agencies and searching for formal listings in the *Monthly Catalog*, comprehensive coverage of grant information can be achieved.

Another format in which grant information is stored is the machine-readable data base. Government data bases containing research-in-progress listings also contain descriptions of federally supported grants and projects. These data bases, examples of which are the National Science Foundation Data Base, and CANCERPROJ of the National Cancer Institute, are discussed in chapter 11 on data bases.

Although some sources list grant information in particular subject areas or for particular agencies, others list them in numerous disciplines and in numerous agencies. One such source is the *Catalog of Federal Domestic Assistance.*[1] The catalog lists programs, projects, services, and activities providing benefits or assistance to the American public. The federal domestic assistance programs included in the catalog are those that may be applied for by state and local governments; United States territories and possessions; private, public, and quasi-public organizations; specialized groups; and individuals.

One of the sixteen categories of assistance included in the catalog is that of Project Grants. This category incorporates listings for fellowships, scholarships, and traineeships, as well as the following types of grants: research, training, experimental and demonstration, evaluation, planning, technical assistance, survey, and construction. Two other pertinent categories in the catalog are Training (instructional programs for non-federal government employees), and Research Contracts (assistance in which funds are handled through contracts rather than grants). Information for each assistance program cited in the catalog includes:

1. Agency administering the program.
2. Goals and objectives of the program.
3. Types of assistance offered.
4. Eligibility requirements.

5. Application and award process.
6. Examples of previously funded activities.
7. Regulations, guidelines, and other literature about the program.
8. Information contacts.

The *Catalog of Federal Domestic Assistance* is published annually, usually in May, and is updated periodically. It is available both in print format and on machine-readable magnetic tape that can be purchased from the National Technical Information Service. The magnetic tape format reflects all textual material published in the program descriptions section of the catalog. For comprehensive information, the catalog should be supplemented by a search of the more current issues of the *Federal Register*, a regulatory publication issued by the Office of Federal Register which is a part of the National Archives and Records Service.

Another publication of general interest is *Commerce Business Daily* (C57.20:(date)) which includes government procurement invitations, contract awards, and notices of federal agencies interested in specific research and development programs.

An example of a publication that describes specific agency programs is *Guide to Programs* of the National Science Foundation (NS1.20:P94/(yr.)). Available free from the foundation, it provides a summary information about various NSF programs. It describes each program in terms of its characteristics and purpose and lists eligibility requirements, closing dates, and addresses for obtaining additional information. Other information such as proposal deadlines can be located in *NSF Bulletin* (NS1.3:(v.nos.&nos.)). An example of a brochure describing a specific program is the foundation's *Program Announcement: Information Science Research* (NS1.2:P94/7).

Federal agencies publish, at varying intervals, lists of research grants awarded by them. Typically, these lists contain information such as the principal investigator, performing organization, title of the project and the amount of grant. National Science Foundation, for example, issues *Daily Congressional Notification of Grants and Contracts Awarded* (NS1.34:(date)). Grants and contracts are arranged in this notification by state and congressional districts. Information on each grant includes name of the organization, individual department, award number, title of the project, duration and amount of grant.

NSF also issues an annual publication, *National Science Foundation Grants and Awards* (NS1.10/4:(yr.), 1963/64-). This publication lists grants by subjects and programs. Under each subject, grants are listed by state and then by institution. Another federal publication listing grants and awards is *Research Awards Index* (HE20.3013/2:(yr.)). It is a source of information on health research currently conducted by non-federal institutions and supported by the health agencies of the Department of Health and Human Services.

In the first volume of the index research programs are listed under about seven thousand subject headings. Under each heading, the titles of the programs are listed in project-number order. The second volume of the index contains Project Identification Data, research contracts, and an alphabetical list of investigators.

The National Referral Center has published a free brochure titled *Selected Information Resources on Scholarships, Fellowships, Grants and Loans.* It describes government, private, and special interest organizations that offer financial assistance to students and scholars. A copy of this brochure can be obtained from: National Referral Center, Science and Technology Division, Library of Congress, 10 First Street, SE, Washington, DC 20540; (202)287-5670.

The National Science Foundation data base, available through NTIS, offers a computer-searchable listing of projects officially completed during the fiscal year. The data base contains technical summaries for research grants; grants for symposia, workshops, conferences, and studies; and science education grants.

A free bibliography in the Superintendent of Documents Subject Bibliography series is SB-258, *Grants and Awards*. It describes publications that are for sale from the government, gives ordering information and prices, and also provides a Superintendent of Documents classification number which can be used to find the publications in many depository libraries. Series SB-258 may be requested from: U.S. Government Printing Office, Superintendent of Documents, Washington, DC 20402.

The list that follows includes government sources of information on grants, awards, fellowships, and scholarships in science and technology.

Aging

National Institutes of Health. National Institute on Aging. *Research Training Opportunities at the National Institute on Aging, Gerontology Research Center* (NIH pub. no. 80-482). Bethesda, MD: National Institute on Aging, 1976- . (HE20.3852:G31/2).

National Institute on Aging. Program Analysis Office. *Extramural Research and Training Program Grant and Contract Summaries: National Institute on Aging FY 1974-1977* (DHEW pub. no. NIH 80-235). Bethesda, MD: National Institute on Aging, 1979. (HE20.3852:EX8).

National Institute on Aging. Program Analysis Office. *Index of Current Research Grants and Contracts Administered by the National Institute on Aging* (DHEW pub. no. NIH 80-1693). Washington, DC: National Institutes of Health, 1977- . (HE23.3002:R31/3).

Aeronautics

National Aeronautics and Space Administration. *NASA Grant and Cooperative Agreement Handbook*. Washington, DC: U.S. Government Printing Office, 1981. Irregular. (NAS1.18:G76/4).

Children

Department of Health & Human Services. Administration for Children, Youth, and Families. *Obtaining Grants and Contracts from the Administration for Children, Youth, and Families* (DHHS pub. no. OHDS 79-30227). Washington, DC: Administration for Children, Youth, and Families, 1977- . (HE23.1002:G76).

National Institutes of Health. National Institute of Child Health and Human Development. *Research Programs of the National Institute of Child Health and Human Development* (NIH pub. no. 81-83). Bethesda, MD: National Institute of Child Health and Human Development, 1970- . (HE20.3352:R31/3/981).

Dentistry

National Institute of Dental Research. *National Institute of Dental Research Programs* (DHHS pub. no. NIH 80-547). Bethesda, MD: The Institute, 1972/73- . Annual. (HE20.3402:D43/5/979).

National Institute of Dental Research. *Trainees and Fellows Supported by the National Institute of Dental Research and Trained during Fiscal Year.* Bethesda, MD: National Institute of Dental Research, 1968- . (HE20.3410:(yr.)).

Energy

Department of Energy. Office of Energy Research. *Instruction and Information on Used Energy-Related Laboratory Equipment Grants for Educational Institutions of Higher Learning* (DOE/ER-0042). Springfield, VA: National Technical Information Service, 1980. (E1.19:0042).

Environment

U.S. Environmental Protection Agency. *EPA and the Academic Community, Partners in Research, Solicitation for Grant Proposals* (EPA 600/8-80-010). Washington, DC: Environmental Protection Agency, 1981. (EP1.2:Ac1/2/981).

U.S. Environmental Protection Agency. *Federal Financial Assistance for Pollution Prevention and Control.* Washington, DC: U.S. Government Printing Office, 1979. (EP1.2:F49).

U.S. Environmental Protection Agency. Grants Administration Division. *EPA Monthly Awards for Construction Grants for Wastewater Treatment Works.* Washington, DC: U.S. Government Printing Office, 1974- . Monthly. (EP1.56:(nos.)).

U.S. Environmental Protection Agency. Grants Administration Division. *Research, Demonstration, Training, and Fellowship Awards.* Washington, DC: Environmental Protection Agency, 1977- . Semiannual. (EP1.35:(nos.)).

U.S. Environmental Protection Agency. Grants Administration Division. *Sources of Information on EPA Grants Awarded, Reports Generated by Grants, and Other Grants-Related Materials.* Washington, DC: Environmental Protection Agency, 1976. (EP1.8:G76).

U.S. Environmental Protection Agency. Grants Administration Division. *State and Local Grant Awards.* Washington, DC: Environmental Protection Agency, 1978- . Semiannual. (EP1.35/2:(nos.)).

U.S. Environmental Protection Agency. Grants Administration Division. *State Priority Lists for Construction Grants for Wastewater Treatment Works* (EPA GAD 4-80-01). Washington, DC: U.S. Government Printing Office, 1980. Irregular. (EP1.56/4:980).

U.S. Environmental Protection Agency. Office of Water and Waste Management. *Handbook of Procedures: Construction Grants Program for Municipal Wastewater Treatment Works* (MCD-03). Washington, DC: Environmental Protection Agency, 1970- . (EP2.28:03/2).

National Institute of Environmental Health Sciences. *National Institute of Environmental Health Sciences Programs* (DHEW pub. no. NIH 80-1905). Washington, DC: U.S. Government Printing Office, 1978-79. (HE20.3552:En8/(date)).

National Institute of Environmental Health Sciences. *Research and Training Programs of the National Institute of Environmental Health Sciences* (DHEW pub. no. NIH 80-1980). Bethesda, MD: National Institutes of Health, 1979. (HE20.3552:R31).

General Science

U.S. Department of Commerce. *Science and Technology Fellowship Program.* Washington, DC: Department of Commerce, 1979/80. (C1.2: Sci2/4/979-80).

Department of Defense. *Basic Research Program.* Washington, DC: U.S. Department of Defense, 1980. (D1.2:R31/9).

Executive Office of the President. Office of Management and Budget. *Catalog of Federal Domestic Assistance.* Washington, DC: U.S. Government Printing Office, 1971- . Annual. (PrEx2.20:(yr.)).

Library of Congress. National Referral Center. Science and Technology Division. *Selected Information Resources on Scholarships, Fellowships, Grants and Loans* (SL77-1). Washington, DC: National Referral Center, 1979. (LC33.11:77-1).

National Aeronautics and Space Administration. Office of University Affairs. *The NASA University Program: A Guide to Policies and Procedures.* Washington, DC: NASA, 1980. (NAS1.18:N21/2).

National Bureau of Standards. *Postdoctoral Research Associateships: Opportunities for Research at the National Bureau of Standards.* Washington, DC: National Bureau of Standards, 1980. (C13.2:R31/6/980).

National Bureau of Standards. *Postdoctoral Research Associateships: Opportunities for Research at the U.S. Department of Commerce, National Bureau of Standards, Washington, D.C. and Boulder, Colorado.* Washington, DC: National Bureau of Standards, 1981. (C13.2:R31/6/981).

National Science Foundation. *Federal Funds for Research and Development.* Washington, DC: U.S. Government Printing Office, 1952- . Annual. (NS1.18:27).

National Science Foundation. *Guidelines for Preparation of Unsolicited Proposals for the Access Improvement Program.* Washington, DC: National Science Foundation, 1975. (NS2.2:P94).

National Science Foundation. *NSF Grant Policy Manual.* Washington, DC: U.S. Government Printing Office, 1979. (NS1.20:G76/2/979).

National Science Foundation. *National Science Foundation Grants and Awards.* Washington, DC: U.S. Government Printing Office, 1963/64- . Annual. (NS1.10/4).

National Science Foundation. Division of Scientific Personnel Improvement. *Student-Originated Studies: Guide for Preparation of Proposals and Project Operation.* Washington, DC: National Science Foundation, 1981. (NS1.20/2:SE-80-22).

Smithsonian Institution. Office of Fellowships and Grants. *Smithsonian Opportunities for Research and Study in History, Art, Science.* Washington, DC: Government Printing Office, 1972- . Irregular. (SI1.2:R31/(yr.)).

U.S. Superintendent of Documents. *Grants and Awards* (Subject Bibliography; SB-258). Washington, DC: U.S. Government Printing Office, 1975- . Irregular. (GP3.22/2:258/7).

Medicine and Health

U.S. Department of Health and Human Services. *Toxicology Research Projects Directory.* Springfield, VA: National Technical Information Service, 1976- . Quarterly. (HE1.48:(v.nos.&nos.)).

Epidemiology Research Projects Directory. Springfield, VA: National Technical Information Service, 1980- . Annual. (Y3.R26/2:Ep4/v.1).

National Institutes of Health. *Grants for Training, Construction, Cancer Control, Medical Libraries* (NIH pub. no. 81-1043). Washington, DC: U.S. Government Printing Office, 1977/78- . Annual. (HE20.3013/5:(yr.)).

National Institutes of Health. *NIH Guide for Grants and Contracts.* Bethesda, MD: NIH, 1971- . Irregular. (HE20.3008/2:(v.nos.&nos.)).

National Institutes of Health. *National Library of Medicine Grant Programs* (DHEW pub. no. NIH 80-260). Bethesda, MD: National Institutes of Health, 1979. (HE20.3602:G76/2/979).

National Institutes of Health. *Research Awards Index* (DHEW pub. no. NIH 81-200). Washington, DC: U.S. Government Printing Office, 1961- . Annual. (HE20.3013/2:(yr./v.nos.)).

National Institutes of Health. *Research Grants* (NIH pub. no. 81-1042). Bethesda, MD: NIH, 1968- . Annual. (HE20.3013/2-2:(yr.)).

National Institutes of Health. *Subject Index of Extramural Research Administered by the National Cancer Institute* (DHHS pub. no. NIH 80-541). Bethesda, MD: National Institutes of Health, 1977/78- . Annual. (HE20.3152:R31/5/(yr.)).

John E. Fogarty International Center for Advanced Study in the Health Sciences. International Cooperation and Geographic Studies Branch. *National Institutes of Health International Awards for Biomedical Research and Research Training* (DHEW pub. no. NIH 80-63). Bethesda, MD: National Institutes of Health, 1969- . Annual. (HE20.3709/2:(yr.)).

National Cancer Institute. Division of Cancer Research Resources and Centers. Grants Financial and Data Analysis Branch. *NCI Grants Awarded by State, City, Institution, and Grant Number.* Bethesda, MD: National Institutes of Health, 1977/78- . Annual. (HE20.3176:(yr.)).

National Heart, Lung, and Blood Institute. *National Heart, Lung, and Blood Institute Program Project Grant; Preparation of the Application* (DHEW pub. no. NIH 79-1005). Bethesda, MD: National Heart, Lung, and Blood Institute, 1979- . Annual. (HE20.3208:P94/(yr.)).

National Heart, Lung, and Blood Institute. *Subject Index of Current Research Grants and Contracts Administered by the National Heart, Lung, and Blood Institute* (DHHS pub. no. NIH 80-705). Bethesda, MD: National Heart, Lung, and Blood Institute, 1975- . Irregular. (HE20.3212:(yr.)).

National Institute of General Medical Sciences. *Pharmacology Research Associate Program of the National Institute of General Medical Sciences, National Institutes of Health* (DHEW pub. no. NIH 81-135). Bethesda, MD: National Institute of Health, 1980- . Annual. (HE20.3452:P49/(yr.)).

National Institute of Neurological and Communicative Disorders and Stroke. *The NINCDS Research Program: Amyotrophic Lateral Sclerosis* (DHHS pub. no. NIH 80-1934). Bethesda, MD: National Institutes of Health, 1980- . Annual. (HE20.3502:Am9/2/(yr.)).

National Institute of Neurological and Communicative Disorders and Stroke. *The NINCDS Research Program: Epilepsy* (DHHS pub. no. NIH 80-1612). Bethesda, MD: National Institutes of Health, 1980- . Annual. (HE20.3502:Ep4/2/(yr.)).

National Institute of Neurological and Communicative Disorders and Stroke. *The NINCDS Research Program: Huntington's Disease* (DHHS pub. no. NIH 80-1935). Bethesda, MD: National Institutes of Health, 1980- . Annual. (HE20.3502:H92/(yr.)).

National Institute of Neurological and Communicative Disorders and Stroke. *The NINCDS Research Program: Multiple Sclerosis* (DHHS pub. no. NIH 81-1614). Bethesda, MD: National Institutes of Health, 1980- . Annual. (HE20.3502:M91/(yr.)).

National Institute of Neurological and Communicative Disorders and Stroke. *The NINCDS Research Program: Muscular Dystrophy and Other Neuromuscular Disorders* (DHHS pub. no. NIH 80-1615). Bethesda, MD: National Institutes of Health, 1980- . Annual. (HE20.3502:M97/2/(yr.)).

National Institute of Neurological and Communicative Disorders and Stroke. *The NINCDS Research Program: Parkinson's Disease* (DHHS pub. no. NIH 80-1616). Bethesda, MD: National Institutes of Health, 1976- . Annual. (HE20.3502:P22/(yr.)).

National Institute of Neurological and Communicative Disorders and Stroke. *The NINCDS Research Program: Spinal Cord Injury and Nervous System Trauma* (DHHS pub. no. NIH 81-1617). Bethesda, MD: National Institutes of Health, 1980- . Annual. (HE20.3502:Sp4/2/(yr.)).

National Institute of Neurological and Communicative Disorders and Stroke. *The NINCDS Research Program: Spina Bifida and Neural Tube Defects* (DHHS pub. no. NIH 80-2135). Bethesda, MD: National Institutes of Health, 1980- . Annual. (HE20.3502:Sp4/3).

National Institute of Neurological and Communicative Disorders and Stroke. *The NINCDS Research Program: Stroke* (DHHS pub. no. NIH 81-1618). Bethesda, MD: National Institutes of Health, 1980- . Annual. (HE220.3502:St8/(yr.)).

National Institute of Neurological and Communicative Disorders and Stroke. *NINCDS Index to Research Grants and Contracts.* Bethesda, MD: National Institutes of Health, 1981- . Annual. (HE20.3516:(yr.)).

National Institutes of Health. Division of Research Grants. Statistics and Analysis Branch. *National Institutes of Health Research Grants* (DHEW pub. no. NIH 80-1042). Bethesda, MD: NIH, 1968- . Annual. (HE20. 3013/2-2/(yr.)).

National Institutes of Health. Division of Research Grants. Statistics and Analysis Branch. *National Institutes of Health Research and Development Contracts* (DHEW pub. no. 80-1044). Washington, DC: U.S. Government Printing Office, 1979- . Annual. (HE20.3013/4:(yr.)).

U.S. Public Health Service. *Profiles of Financial Assistance Programs* (DHHS pub. no. OASH 80-50,002). Washington, DC: U.S. Government Printing Office, 1980. (HE20.2:F49/980).

Health Resources Administration. Bureau of Health Manpower. *BHM Support: Directory of Grants, Awards, and Loans* (DHEW pub. no. HRA 80-23). Rockville, MD: Public Health Service, 1978- . Annual. (HE20.6612:(yr.)).

Health Resources Administration. Bureau of Health Manpower. Program Management Information Systems Branch. *Trends in BHM Program Statistics; Grants, Awards, Loans* (DHEW pub. no. HRA 80-26). Washington, DC: U.S. Government Printing Office, 1957/75- . Irregular. (HE20.6602:P94/(yrs.)).

Mental Health

Public Health Service. Alcohol, Drug Abuse, and Mental Health Administration. *Alcohol, Drug Abuse, and Mental Health Administration Research Programs* (DHHS pub. no. ADM 81-942). Rockville, MD: U.S. Department of Health and Human Services, 1973- . (HE20.8002:R31).

National Institute of Mental Health. Division of Biometry and Epidemiology. *Grant Program in Mental Health Service System Research.* Rockville, MD: National Institute of Mental Health, 1979. (HE20.8102: G76/3).

National Institute of Mental Health. Division of Extramural Research Programs. Program Analysis and Evaluation Section. *Alcohol, Drug Abuse, Mental Health, Research Grant Awards* (DHHS pub. no. ADM 81-319). Rockville, MD: National Institute of Mental Health, 1972/73- . Annual. (HE20.8118:(yr.)).

Transportation

U.S. Federal Highway Administration. Office of Development, Implementation Division. *FHWA Research and Development Implementation Catalog*. Washington, DC: Federal Highway Administration, 1975- . (TD2.36/3:(yr.)).

U.S. Federal Highway Administration. *Federally Coordinated Program of Highway Research and Development*. Washington, DC: U.S. Government Printing Office, 1974/75- . Annual. (TD2.42:(yr.)).

U.S. Federal Highway Administration. Office of Research and Development. *Federally Coordinated Program of Research and Development in Highway Transportation*. Washington, DC: Department of Transportation. (TD2.2:R31/2/v.1-5/980).

National Highway Institute. *Education and Training Information Exchange Bulletin*. Washington, DC: National Highway Institute, 1973- . Monthly. (TD2.39:(no.)).

U.S. Urban Mass Transportation Administration. *Directory of Research, Development & Demonstration Projects* (UMTA-MA-06-0086-81-1). Washington, DC: Government Printing Office, 1969- . Annual. (TD7.9:(yr.)).

Others

U.S. Fire Administration. *USFA Fellowship Program*. Washington, DC: U.S. Fire Administration, 1980. (FEM1.102:Un3).

National Fire Safety and Research Office. *Catalog of Grants, Contracts and Interagency Transfers*. Washington, DC: National Fire Safety and Research Office, 1976-77. (C58.12:(date)).

National Aeronautics and Space Administration. *Crustal Dynamics and Earthquake Research*. Washington, DC: NASA, 1980. (NAS1.53: OSTA80-2).

U.S. National Marine Fisheries Service. *Grant-in-Aid for Fisheries Program Activities* (NOAA S/T 80-161). Washington, DC: National Marine Fisheries Service, 1979- . Annual. (C55.323:(yr.)).

REFERENCES

1. Executive Office of the President, Office of Management and Budget, *Catalog of Federal Domestic Assistance* (Washington, DC: U.S. Government Printing Office, 1981).

3

RESEARCH IN PROGRESS*

A research project, once initiated after obtaining support in the form of a grant or a contract, may take two or more years to be completed. After the project is initiated, it may take an additional year or two before the research scientist starts presenting the information generated from the project through conference papers, technical reports, and journal articles. That is, until the scientist starts releasing the information, other scientists may not be aware of its existence and its progress.

This presents the possibility of an unnecessary duplication of research on the part of other scientists. Knowledge of various projects in progress is necessary for a scientist who is embarking on a new project since that knowledge gives an excellent idea of the state-of-the-art of research in his or her field of interest. In this connection, the federal government actively disseminates information on research projects in progress.

To monitor research projects about to be initiated, those currently in progress, and those recently completed, the scientist or researcher relies on ongoing research information systems. These systems seek to bridge the prepublication gap by providing information on research projects between their inception and the time when research reports or intermediate publications appear in the open literature.

---

*Parts of this chapter previously appeared as: "Federal Government Research in Progress: Information for Science and Technology," *Database* 6 (February 1983): 36-43.

Ongoing research information systems seek to help avoid undesired duplication of research efforts, to encourage maximum utilization of research and development funds, and to provide national coordination and planning of future research.[1] Such systems endeavor to answer questions such as

What research is being performed

Where

Who is performing it

How has it been funded

When did it begin, and what is the anticipated completion date

Where can additional information on this project be obtained[2]

Because of the need for manipulation of these data, in addition to requirements of currency and for weeding records every two or three years, these systems are typically computerized. Frequently the data recorded in these systems can be retrieved only through computer searches. In other cases, the data are available in print formats such as journals or serials.

The United States has been a pioneer in the development of ongoing research information systems, beginning in the 1950s with efforts to provide access to ongoing medical research. These efforts led to the development of the Smithsonian Science Information Exchange, a centralized national system monitoring basic and applied research in all areas of science.

THE SMITHSONIAN SCIENCE INFORMATION EXCHANGE

The Smithsonian Science Information Exchange (SSIE) was phased out on October 30, 1981, a victim of federal budget cuts. Before then, the SSIE was the national registry for research in progress in the sciences in the United States. As the only research-in-progress information system offering national coverage of all areas of science, it was a vital link for identifying elusive ongoing research.

The exchange provided interdisciplinary national and international coverage of current science research by encouraging government agencies and private organizations pursuing or supporting research to voluntarily submit current research summaries. These summaries made up the research project data base for SSIE.

The types of agencies and organizations contributing to this data file were diverse, including federal government agencies, state and local governments, foundations, associations, private researchers, academic institutions, and, on a more limited scale, private industry and foreign organizations.

A portion of the former services of SSIE is assumed by the National Technical Information Service (NTIS). NTIS has focused on the SSIE research project data base, updating it monthly using data tapes supplied by federal government agencies. The NTIS research project data base can be leased from NTIS, and is also searchable through the three commercial data base systems: Bibliographic Retrieval Services (BRS) (1200 Route 7, Latham, NY 12110); DIALOG® Information Retrieval Services (DIALOG) (3460 Hillview Avenue,

Palo Alto, CA 94304); and ORBIT® Information Retrieval System (ORBIT) (2500 Colorado Avenue, Santa Monica, CA 90406).

Agencies submitting research-in-progress data to NTIS include the National Institutes of Health; the Departments of Agriculture, Energy, and Defense; the National Bureau of Standards; the U.S. Geological Survey; and NASA. Some agencies never maintained their own research-in-progress data files, depending upon SSIE to generate special subject files from the SSIE current research summary file. These agencies are not currently being served by NTIS, since NTIS is inputting only from agency-generated data tapes. The National Science Foundation is currently in this predicament, although NTIS plans to input from hard copy in the future.

SSIE print products have been discontinued. These include Research Information Packages, which were summaries of current research activities on high interest topics, and catalogs of ongoing research. One such catalog, published by SSIE, is still available: *Information Services on Research in Progress: A Worldwide Inventory*, a major directory of worldwide ongoing research information systems (NS1.2:In2/7).[3] This source lists 179 ongoing research information systems in planning, pilot, or operational status, located in 53 countries. Included are national, subnational, international, and regional systems.

RESEARCH-IN-PROGRESS INFORMATION SYSTEMS

The following listing provides details about selected research-in-progress information systems in the United States. The systems listed reflect the two common patterns of control of ongoing research on a national basis. The first is the single, national system which aims at comprehensive national coverage; the second involves separate systems developed in specific subject areas or programs of interest.[4]

In the United States, a number of the special subject information systems have used the Smithsonian Science Information Exchange data base to generate special listings on research in progress in their subject areas. Others have neither published their own listings nor created special data files, relying instead on the direct services of SSIE. In the latter case, the agencies periodically submitted notices of new and updated research projects to SSIE for cataloging and dissemination. The impact of the dissolution of SSIE upon these approaches is yet to be determined.

Title: Aquaculture
Sponsor: National Oceanographic Data Center and National Oceanic and Atmospheric Administration
Scope: A computer data base which includes listings for current research projects related to growing freshwater, brackish or marine plants and animals
Availability: Access to this data base is through ENDEX/OASIS services of NOAA.

Title: CANCERPROJ (Cancer Research Projects)
Sponsor: National Cancer Institute, National Institutes of Health, U.S. Department of Health and Human Services
Scope: Private and federally funded cancer research in U.S. and other countries
Availability: This machine-readable data file can be searched at terminals with online access to MEDLARS. CANCERPROJ is part of the CANCERLINE data base of the National Library of Medicine.

Title: Clearinghouse for On-Going Research in Cancer Epidemiology
Sponsor: International Cancer Research Data Bank Program (ICRDB), the International Agency for Research on Cancer (Lyon, France), and the German Cancer Research Center (Heidelberg)
Scope: The Clearinghouse collects and disseminates information on cancer epidemiology and human cancer causation research, wordwide.
Availability: ICRDB Program Office, National Cancer Institute, Blair Bldg., Room 114, 8300 Colesville Road, Silver Spring, MD 20910
Companion Resources: Directory of On-Going Research in Cancer Epidemiology (annual), published by ICRDB

Title: Computer Retrieval of Information on Scientific Projects (CRISP)
Sponsor: National Institutes of Health, Public Health Service, U.S. Department of Health and Human Services
Scope: A computer-based information retrieval system for research grants and contracts programs of the Public Health Service, and those conducted intramurally by the National Institutes of Health and the National Institute of Mental Health. The system offers the special capability to subdivide projects into their individual research components, providing detailed information on large grants.
Availability: Free of charge to all Public Health Service grantee institutions from the Research Documentation Section, Statistics and Analysis Branch, Division of Research Grants, National Institutes of Health, Bethesda, MD 20205; (301)496-7543
Companion Resources: Projects are indexed according to the *Medical and Health Related Sciences Thesaurus* (*MHRS*).[5]

Title: Current Research Information System (CRIS)
Sponsor: Science and Education, U.S. Department of Agriculture
Scope: A computerized information storage and retrieval system for current agricultural and forestry research activities of the U.S. Department of Agriculture and its agencies, the State Agriculture Experiment Stations, and other cooperating institutions
Availability: Searches of the current file on a demand basis from USDA, Science and Education, National Agricultural Library Bldg., Beltsville, MD 20705; (301)344-3846. CRIS is also available online through DIALOG (CRIS/USDA Online).

Companion Resources: Ongoing research information from CRIS is included in the *Inventory of Agriculture*, published by USDA annually.[6] Ongoing research project information recorded in CRIS is forwarded for inclusion in the NTIS research project data file.

Title: Dental Research Data Officer
Sponsor: National Institute of Dental Health, National Institutes of Health
Scope: Dental sciences research in progress nationally, primarily that sponsored by the National Institute of Dental Research, plus limited information from other organizations worldwide
Availability: On-demand searches and statistical analysis through National Institute of Dental Research, National Institutes of Health, Bethesda, MD 20014; (301)496-7720
Companion Resource: Dental Research in the United States and Other Countries published annually by the National Institute of Dental Research[7]

Title: Dental Research in the United States and Other Countries[8]
Sponsor: National Institute of Dental Research, National Institutes of Health
Scope: An annual catalog published in cooperation with SSIE, and listed ongoing research projects of interest to dentistry from all over the world. This publication included all dental research projects registered with the Smithsonian Science Information Exchange.
Availability: Available for purchase from the Superintendent of Documents; single free copy available from Dental Research Data Officer, National Institute of Dental Research, National Institutes of Health, Bethesda, MD 20014; (301)496-7220. Available to depository libraries (HE20.3402:D43/978).

Title: Directory of On-Going Research in Cancer Epidemiology[9]
Sponsor: This directory is a joint project of the International Agency for Research on Cancer (Lyon, France), and the German Cancer Research Center (Heidelberg) within the framework of the International Cancer Research Data Bank (ICRDB) program of the National Cancer Institute.
Scope: Summaries of 1,313 projects from eighty countries, each focusing on the relationship between many aspects of the environment and human cancer
Availability: For sale from the National Technical Information Service
Companion Resources: CANCERPROJ data file (NLM); *Special Listings, Cancergrams*, and *Oncology Overviews* from ICRDB

Title: Directory of On-Going Research in Smoking and Health[10]
Sponsor: Technical Information Center, Office on Smoking and Health, Public Health Service, U.S. Department of Health and Human Services

Scope: Current research on smoking, tobacco, and tobacco use, worldwide

Availability: Published biennially and for sale by the Superintendent of Documents; free copies may be obtained by writing to Technical Information Center, Office on Smoking and Health, Park Bldg., Room 116, 5600 Fishers Lane, Rockville, MD 20857. Available to depository libraries (HE20.7015:(yr.)).

Title: Directory of Solar Energy Research Activities in the United States[11]

Sponsor: U.S. Department of Energy

Scope: A compilation of solar energy research projects generated from the data files of the Smithsonian Science Information Exchange. The first edition of this directory (1980) lists projects in progress by October 1977 or later.

Availability: Annual updates of this directory are planned. Single free copies can be obtained from Document Distribution Service, ATTN: Research Directory, Solar Energy Research Institute, 1617 Cole Blvd., Golden, CO 80401; or purchase from National Technical Information Service. Available to depository libraries (E1.28:SERI/SP-644-690).

Title: DOE Research in Progress (RIP)

Sponsor: U.S. Department of Energy

Scope: A data base that describes new and ongoing energy and energy-related research projects carried out by or sponsored by DOE

Availability: Available to DOE personnel and contractors. Write to Technical Information Center, P.O. Box 62, Oak Ridge, TN 37830; (615)576-1303. Will also be computer searchable through NTIS in 1983.

Title: Highway Research Information Service (HRIS)

Sponsor: Transportation Research Board of the National Research Council and U.S. Federal Highway Administration

Scope: A computerized system offering summaries of U.S. and some foreign research on highways, highway transportation, nonrail transit, and urban transportation planning

Availability: Online access, and on-demand searches through the Transportation Research Board, 2101 Constitution Avenue, Washington DC 20418; (202)389-6358

Companion Resources: World Survey of Current Research and Development on Roads and Road Transport, prepared annually by the International Road Federation; *HRIS Abstracts*, published quarterly

Title: Interagency Committee on Population Research

Sponsor: Center for Population Research, National Institutes of Health, U.S. Public Health Service

Scope: Research in the U.S. related to the biological, medical, demographic, economic, political, psychological, historical, cultural, and sociological aspects of population

Availability: Research project information is made available through two annual publications: *Inventory and Analysis of Federal Population Research*,[12] and *Inventory of Private Agency Population Research*.[13] Copies are available from ICPR Chairman, Center for Population Research, National Institutes of Health, U.S. Public Health Services, Landon Bldg., Room A-721, Bethesda, MD 20014; (301)496-1101; both titles are also available to depository libraries.

Title: Maritime Research Information Service (MRIS)

Sponsor: Transportation Research Board and U.S. Maritime Administration

Scope: Maritime transportation research projects, primarily in the U.S., with some international coverage

Availability: On-demand searches through the Transportation Research Board, 2101 Constitution Avenue, Washington, DC 20418; (202)389-6687

Title: "On-Going Research Projects" (aerospace)

Sponsor: National Aeronautics and Space Administration and Smithsonian Science Information Exchange

Scope: The aerospace oriented "On-Going Research Projects" is inserted into each issue of *Scientific and Technical Aerospace Reports* (*STAR*). Includes project announcements for active NASA grants and university contracts, summary portions of recently updated *NASA Research and Technology Objectives and Plans*, and non-NASA research projects. Projects listed are in aeronautics, space research and development, and aerospace aspects of earth resources, energy development, conservation, oceanography, environmental protection, and urban transportation.

Availability: Semimonthly in *STAR*. Note: Discontinued for the present with the demise of SSIE.

Title: Oncology Overviews

Producer: International Cancer Research Data Bank Program (ICRDB) of the National Cancer Institute

Scope: Abstracts of selected publications about high-interest cancer research topics. Over thirty different *Oncology Overview* topics are now available, and several new topics are added yearly.

Availability: For sale from the National Technical Information Service

Title: Psychopharmacology Bulletin

Sponsor: National Institute of Mental Health, U.S. Department of Health and Human Services

Scope: Includes information on research in progress in the U.S. and abroad related to pharmacologic and somatic treatments

Availability: This quarterly publication is available on subscription from the Superintendent of Documents; available to depository libraries (HE20.8109:(v.nos.&nos.)).

Title: Research and Technology Work Unit Information System
Sponsor: Defense Technical Information Center (DTIC)
Scope: Summaries of research in all scientific fields. Includes project summaries from U.S. government agencies, other national organizations, and Canadian organizations for research in the areas of science and technology pertinent to Department of Defense research, development, testing, and evaluation interests.
Availability: Online access, and on-demand searches from Defense Technical Information Center, ATTN: DTIC-DDR, Bldg. 5, Cameron Station, Alexandria, VA 22314; (202)274-7633. Restricted to Defense Technical Information Center registered users.
Companion Resources: DTIC Data Base

Title: Research and Technology Operating Plan Summary (RTOPS)
Sponsor: National Aeronautics and Space Administration
Scope: Summaries of NASA-funded research, both in-house and through contract, grant, or interagency agreement
Availability: This annual publication can be purchased from the National Technical Information Service. Copies are also available for reference in NASA Center libraries, and libraries of most NASA contractor organizations. Also accessible through computer searches of NASA/RECON, and literature searches prepared by NASA's Scientific and Technical Information Branch.
Companion Resources: NASA/RECON Data Base

Title: Research Digest Series
Sponsor: National Center for Health Services Research, Public Health Service, U.S. Department of Health and Human Services
Scope: This publication series includes summaries of ongoing or completed research related to health service programs and providers of health services, sponsored by NCHSR.
Availability: Available from the National Technical Information Service. For a list of titles write: Publications and Information Branch, NCHSR, Room 7-44, 3700 East-West Highway, Hyattsville, MD 20782; (301)436-8970.

Title: Research in Progress (year)
Sponsor: U.S. Army Research Office
Scope: Describes ongoing research supported by the Army Research Office, the Defense Advanced Research Projects Agency, several army commands, and the Defense Nuclear Agency. One volume of this series covers metallurgy, materials science, mechanics, aeronautics, chemistry, and biological sciences. The second volume covers physics, electronics, mathematics, and geosciences.[14]
Availability: Each annually published volume can be purchased from the National Technical Information Service: available to depository libraries.

Title: Smithsonian Science Information Exchange (SSIE) Data File
Sponsor: Since 1981, maintained by the National Technical Information Service
Scope: Ongoing research project, information submitted by government agencies
Availability: Online access through BRS, ORBIT, and DIALOG; available for lease from NTIS

Title: Special Listings of Current Cancer Research
Sponsor: International Cancer Research Data Bank Program (ICRDB) of the National Cancer Institute
Scope: Annual compilations of current research activity in 55 cancer research areas, in over 83 countries. There are about 55 individual Special Listings titles, each covering a different major area of cancer research, published yearly.
Availability: For sale from the National Technical Information Service. Lists of available titles can be obtained by writing ICRDB Program, National Cancer Institute, Westwood Bldg., Room 10A 18, Bethesda, MD 20205. Qualified cancer researchers may register their research projects with the ICRDB and receive the *Special Listings* in their research area free of charge; also available to depository libraries (HE20.3173:(nos.)).
Companion Resources: Directory of Cancer Research Information Resources, which lists cancer information sources in published literature, computer-based information systems, special collections, audiovisuals, research project information resources, dial-access services, and cancer registries. A complimentary copy of this publication may be obtained from ICRDB.

Title: Toxicology Information Program (TIP)
Sponsor: National Library of Medicine
Scope: U.S., and some international, toxicology research and related data on biomedical production and use
Availability: Research project information is disseminated through two publications: TOX-TIPS (monthly), (HE20.3620:(nos.)), and *Toxicology Research Projects Directory* (quarterly, HE1.48:(v.nos. &nos.)). The National Technical Information Service, 5285 Port Royal Road, Springfield, VA 22161. Also searchable through TOXLINE.
Companion Resources: TOXLINE data base, which contains a subfile of toxicology/epidemiology research projects (RPROJ) from the Smithsonian Science Information Exchange; *Toxicology Research Projects Directory* (quarterly)[15]

Title: Water Resources Scientific Information Center
Sponsor: Office of Water Research and Technology, U.S. Department of the Interior
Scope: Research on water resources and physical, engineering, social, legal aspects of water resources, primarily U.S., with some foreign coverage

Availability: Online access through regional Water Resources Scientific Information Centers.

REFERENCES

1. David F. Hersey, "Information Systems for Research in Progress," in Martha E. Williams, ed., *Annual Review of Information Science and Technology*, Vol. 13 (White Plains, NY: Knowledge Industry Publications, Inc., 1978), p. 263.
2. Smithsonian Science Information Exchange, *Information Services on Research in Progress: A Worldwide Inventory* (Springfield, VA: National Technical Information Service, 1978), p. 4 (PB-282 025) (NS1.2:In2/7).
3. Ibid.
4. Ibid., p. 6.
5. U.S. National Institutes of Health, *Medical and Health Related Science Thesaurus* (DHEW pub. no. NIH 79-199) (Rockville, MD: Department of Health, Education, and Welfare, Public Health Service, National Institutes of Health, 1979) (HE20.3023:979).
6. U.S. Technical Information Systems, *Inventory of Agricultural Research* (Agricultural Reviews and Manuals Headquarters Series) (Beltsville, MD: Department of Agriculture, Science and Education Administration, Technical Information Systems, 1977) (A106.12: H-4/v.1; A 106.12:H-5).
7. U.S. Department of Health and Human Services, Public Health Service, National Institutes of Health, *Dental Research in the United States and Other Countries: A Catalog of Dental Research Projects Sponsored during Fiscal Year 1977 by Federal and Non-Federal Organizations* (NIH pub. no. 81-450) (Washington, DC: U.S. Government Printing Office, 1970- . Annual) (HE20.3402: D43/978).
8. Ibid.
9. C. S. Muir and G. Wagner, *Directory of On-Going Research in Cancer Epidemiology* (IARC pub. no. 38) (Lyon, France: International Agency for Research on Cancer, 1981).
10. U.S. Department of Health and Human Services, Office on Smoking and Health, Public Health Service, *1980 Directory, On-Going Research in Smoking and Health* (Washington, DC: U.S. Government Printing Office, 1967- . Annual) (HE20.7015:(yr.)).
11. U.S. Department of Energy, Solar Energy Research Institute, Academic and University Programs Branch, *Directory of Solar Energy Research Activities in the United States* (Springfield, VA: National Technical Information Service, 1980) (E1.28:SERI/SP-644-690).
12. U.S. Interagency Committee on Population Research, *Inventory and Analysis of Federal Population Research* (DHHS pub. no. NIH 80-2238) (Washington, DC: National Institutes of Health, 1969- . Annual) (HE20.3362/2:(yr.)).

13. U.S. Department of Health and Human Services, Public Health Service, National Institutes of Health, *Inventory of Private Agency Population Research* DHHS pub. no. NIH 80-694) (Bethesda, MD: National Institutes of Health, 1978- . Annual) (HE20.3362/2-2:(yr.)).

14. U.S. Army Research Office, *Research in Progress: Chemistry, Biological Sciences, Engineering Sciences, Metallurgy and Materials Science, European Research Program* (Research Triangle Park, NC: Army Research Office, 1979) (D101.52/5-5:979); U.S. Army Research Office, *Research in Progress: Physics, Electronics, Mathematics, Geosciences* (Research Triangle Park, NC: Army Research Office, 1977- . Annual) (D101.52/5-4:(yr.)).

15. U.S. Department of Health, Education, and Welfare, *Toxicology Research Projects Directory* (DHEW pub. no. OS) (Springfield, VA: National Technical Information Service, 1976- . Quarterly) (HE1.48:(v.nos.&nos.)).

4

TECHNICAL REPORTS

INTRODUCTION

Technical reports, which came into prominence at the end of the Second World War,[1] constitute one of the principal media for primary communication in science and technology. They appear under a variety of names such as periodic progress reports (e.g., quarterly progress reports, annual technical reports, and final reports), technical memoranda, technical notes, research notes and scientific reports.

The report literature, presenting unique problems of bibliographic control, have often annoyed librarians and other information professionals.[2] The scientific community, lacking confidence in this so-called "unpublished" literature, have tried to ignore it.[3] In spite of the absence of wholehearted support from these two crucial constituencies, technical reports have proved durable. The growth of technical reports, in fact, has outpaced that of journal articles, patents, dissertations, and books.[4] In terms of the number of items produced per 1,000 scientists or engineers, technical reports showed a substantial rate of growth compared to other forms of technical literature.

Technical reports present "the results of a scientific investigation or a technical development, test or evaluation ... in a form suitable for dissemination to the technological community.... [They] normally contain sufficient data to enable the qualified reader to evaluate the investigative process of the original research and development."[5] This definition, however, is somewhat ambiguous since it applies equally well to other forms of scientific literature such as journal articles, dissertations, and conference papers. What distinguishes the technical reports from other forms of primary scientific literature is: 1) the technical reports are not necessarily addressed to the fellow scientists, and 2) they are not always available freely to the scientific community.

Technical reports are produced not only to communicate scientific information but also to satisfy the requirements of accountability. For example, if a federal agency contracts for or provides a research grant to conduct research and development work, it may require a report or reports detailing the progress of the project from the contractor or grantee. Various agencies of the United States government have been active during and after the Second World War in sponsoring scientific research and development programs through their own laboratories as well as outside organizations. For example, in 1980 federal obligations for R&D were $31.7 billion of which $8 billion were obligations for federal intramural research. The bulk of the remaining monies went to industry ($14.4 billion), universities and colleges ($4.3 billion), and to Federally Funded Research and Development Centers (FFRDCs) and other nonprofit institutions ($4.6 billion).[6]

Under these conditions of highly decentralized research and development environment, technical reports became a facile means of communication between federal agencies sponsoring R&D work and the external agencies performing work for the government. Communication of scientific information to the scientific community, then, is a secondary objective of the technical report.

A consequence of the technical report serving as a communication link between the contracting/monitoring agency and the contractor/grantee organization is that its distribution can be restricted by the parties involved. For example, the federal agency may restrict the distribution of a report for national security reasons while the contractor/grantee may do so to protect proprietary information. This can be exemplified by the fact that in 1977 $15 billion out of federal R&D obligations of $24 billion went to national defense and space research. Both these areas produce information whose distribution is likely to be restricted on the grounds of national security.[7] As has already been pointed out, private industry accounts for $14.4 billion of 1980 federal R&D obligations. Private industry is likely to restrict the distribution of information stemming from R&D work on grounds of its being of a proprietary nature.

In other words, technical reports differ from other forms of scientific literature in two major respects. First, they are not meant to serve as an *orderly* means of scientific communication because the information system represented by them is not operated by scientists for scientists and is not subjected to a prior screening process by them.[8] Second, they are not *open* because they are not always "freely available to anyone who pays the subscription or has access to a library."[9] These differences notwithstanding, technical report literature is firmly entrenched in the scientific and technical communication process.

TECHNICAL REPORTS AND SCIENTIFIC COMMUNICATION

The importance of technical reports is derived from their close connection with the formal scientific information process. First, at least part of the information contained in technical reports eventually finds its way into the archival scientific information. Second, there is a considerable time lag between the preliminary publication of a technical report and the subsequent appearance of information contained in it in a journal article.

Based on a study of 1,082 technical reports whose publication histories could be traced, Gray and Rosenborg estimated that "60-65% of unclassified technical reports contain publishable information" which would "warrant publication in a scientific or technical journal, book or other generally-available source."[10] They further estimate that "for about half of the technical reports that contain publishable data, all such information is published within 2-3 years."[11] Garvey supports this observation by stating that "the main content of almost half of the 'technical reports' produced by scientists end up being published as journal articles."[12]

This process, however, is not precise. For example, information from a technical report may appear in the form of a number of journal articles and vice versa.[13] Another complication is that the authorships and titles of journal articles may not be the same as those of their corresponding technical reports. In any case, this relationship between technical reports and journal articles indicates the importance of scientific and technical information presented in these reports.

Added to the publishability of the scientific information contained in the technical reports is the fact that they are more detailed than the corresponding journal articles. The Committee on Scientific and Technical Information (COSATI)* Task Group on the Role of the Technical Report recognizes the technical report's "customary comprehensive treatment of an application, together with all of the ancillary information" and points out that it is "significant for anyone facing the problem of making use of similar techniques."[14] Similarly, Garvey points out that "any scientist who is especially interested in a journal article would likely benefit from examining its technical report counterpart" for its more detailed and complete information, including "figures and tabular material not appearing in the journal article" and "supplemental material such as diagrams of equipment and photographs of apparatus, specimens, subjects, etc."[15]

Garvey and Griffith further believe that "the individual researcher often finds it easier to reproduce or follow up an experiment described in a technical report than one described in a journal article."[16] Gray and Rosenborg contend that even when a technical report may not contain "publishable information," it nevertheless may contain valuable scientific data.[17] For example, the reports that present "negative results" can be invaluable in deterring researchers from unproductive approaches.[18] Similarly, many reports contain extensive tabulations of data which are useful to the researchers in the field. Editors of journals are typically reluctant to publish negative results and detailed tabulations of data. In other words, technical reports contain valuable scientific information whether this information is "publishable" in the traditional sense or not.[19]

The importance of technical reports is further enhanced by the fact that technical reports, as has already been pointed out earlier, precede their corresponding journal articles in the flow of communication of scientific information. The time lag between the publication of a technical report and that of its corresponding journal article could be at least a year if not longer. For researchers, access to technical reports means access to timely and current information in their areas of interest. Garvey believes that "between a quarter and

*COSATI was a part of the now defunct Federal Council for Science and Technology.

a half of the research scientists have used information gained from a 'technical report' in the course of their own research—information which they later reported in a journal article."[20] The technical report is timely because it is not subjected to delays involved in journal publication. In contrast to journal articles, technical reports are published in near-print formats, have fewer stylistic and content restrictions, and are not usually subjected to time-consuming refereeing processes. Technical reports, in other words, present useful information—information which is publishable and which, in fact, will appear in traditional scientific communication media—much earlier than in media such as journal articles.

The scientific community, however, may not see the technical reports as an unmixed blessing. As far as scientists and editors of scientific journals are concerned, the technical reports are an "informal" domain and are not yet part of formal scientific information.[21] In fact, many journal editors discourage citing technical reports in the articles published in their journals. Even when a journal article contains substantially the same information as that contained in its corresponding technical report, this fact may not be made clear in the journal article. Similarly, even though an author of a scientific paper obtained useful information from a technical report, its journal counterpart may be cited rather than the original technical report. An unfortunate consequence of these policies is the apparent under-utilization of technical report literature.

From the librarians' point of view, technical reports are important enough to be brought under bibliographic control. However, because of their temporal relationship with journal articles, technical reports become obsolescent more rapidly than journal articles. The same relationship further results in the duplication of scientific information and in the undesirable growth of access tools to the scientific information.[22, 23]

BIBLIOGRAPHIC SOURCES FOR TECHNICAL REPORTS

A number of bibliographic sources are helpful in identifying and locating federally sponsored technical reports. These sources can be divided into three groups: 1) general sources, 2) mission- or subject-based sources, and 3) agency bibliographies.

GENERAL SOURCES

Government Reports Announcements & Index (*GRA&I*) (C51-9/3:(v.nos. &nos.)) and the *Monthly Catalog of United States Government Publications* (*Monthly Catalog*) (GP3.8:(date)) fall into this category. Of these, the *GRA&I* is the more important source for technical reports. *GRA&I* is published by the National Technical Information Service (NTIS) and the *Monthly Catalog* is published by the Government Printing Office (GPO).

GRA&I, the major access tool for unclassified, unlimited technical reports, is published every two weeks by the National Technical Information Service. It announces "U.S. and foreign government-sponsored research, development and engineering reports and other analyses prepared by national and local governmental agencies, their contractors or grantees, or by Special Technology Groups." It also includes the reports announced in *Energy Research Abstracts* (*ERA*) (E1.17:(v.nos.&nos.)) and *Scientific and Technical Aerospace Reports* (*STAR*) (NAS1.9/4:(v.nos.&nos.)).

Citations to the reports are arranged in *GRA&I* under twenty-two major subject categories and their numerous subcategories. The twenty-two categories are

1. Aeronautics
2. Agriculture
3. Astronomy and Astrophysics
4. Atmospheric Sciences
5. Behavioral and Social Sciences
6. Biological and Medical Sciences
7. Chemistry
8. Earth Sciences and Oceanography
9. Electronics and Electrical Engineering
10. Energy Conversion (non-propulsive)
11. Materials
12. Mathematical Sciences
13. Mechanical, Industrial, Civil, and Marine Engineering
14. Methods and Equipment
15. Military Sciences
16. Missile Technology
17. Navigations, Communications, Detection, and Countermeasures
18. Nuclear Science and Technology
19. Ordnance
20. Physics
21. Propulsion and Fuels
22. Space Technology

GRA&I has subject, personal author, corporate author, contract number, and accession/report number indexes. *GRA&I*, however, is not comprehensive. The confusion between the roles of the Government Printing Office, NTIS, and other federal agencies with statutory responsibilities for information dissemination are some of the reasons behind this lack of comprehensiveness.

NTIS does not announce some newly declassified reports if, in its judgment, these reports have no sale value. Access to older NTIS reports is improved by *NTIS Library Reference Files 1943-1972*, marketed by Microforms International Marketing Corporation (Fairview Park, Elmsford, NY 10523). The *Files*, available in 16mm reels or cartridges or 35mm reels of microfilm, consists of Report/Accession Number Files 1943-1971, Cumulative Subject Indexes 1943-1971, and Cumulative Personal Author, Corporate Author, and Contract Number Indexes 1964-1971. The Report/Accession Number Files, in turn, consist of PB, AD, N, and A-Z files. PB report number files cover the period 1943-1971 while the other three files cover the years 1964 to 1971. In these files the abstracts and bibliographic data of the reports are filed in a single alphabetical and numerical sequence. Another useful retrospective source is the NTIS publication *NTIS Title Index* (PB82-916098) and its quarterly updates. The basic set includes all reports entered into the NTIS data base from 1964 through 1982. It is issued in microfiche format and is available from NTIS on subscription. The index has three sections: KWOC title index, accession/report number index, and personal author index.

The *Monthly Catalog* also announces numerous technical reports. However, the number of reports announced in this source is much smaller than that announced in *GRA&I. Monthly Catalog* does not attempt to be comprehensive for technical reports, announcing only those that were published by the Government Printing Office or brought to the attention of the *Monthly Catalog* by the federal agencies. Publications in the *Monthly Catalog* are arranged by Superintendent of Documents classification numbers—i.e., broadly speaking, by federal departments and agencies.

Data included on each entry in the *Monthly Catalog* follow Anglo-American cataloging rules. There are author, title, subject, keyword, and series/report indexes to the *Monthly Catalog* entries. Access to older reports announced in the *Monthly Catalog* is provided by Readex Microprint Corporation (101 Fifth Avenue, New York, NY 10002) which issues microprints of nondepository (1953-) and depository items in two series.

A complementary publication to the *Monthly Catalog* is the *CIS/Index*, published by Congressional Information Service. Although, strictly speaking, it is not a source for technical reports, it is a useful source in locating congressional reports and other publications that have a bearing on the federal scientific research and information policies.

MISSION-BASED SOURCES

Technical Abstracts Bulletin (*TAB*), *Scientific and Technical Aerospace Reports* (*STAR*), *Selected Water Resources Abstracts* (I1.94/2:(v.nos.&nos.)), and *Energy Research Abstracts* (*ERA*) are examples of mission- or subject-oriented sources. They do not attempt to announce all technical reports, but limit their coverage by means of a predetermined criteria compatible with the missions of their parent agencies. For example, *TAB* announces reports that record "scientific and technical results of Defense-sponsored research, development, test, and evaluation efforts" submitted by Defense facilities and their contractors. Further, *TAB* includes both "classified and unclassified documents that cannot be released to the general public." In fact, *TAB* and its indexes are currently issued as a single confidential publication.

On the other hand, Defense reports which are unclassified and have no limitations on their distribution are announced in *GRA&I* along with declassified reports. The documents announced in *TAB* are arranged under the same twenty-two subject categories followed by *GRA&I*. However, there is a slight difference in the subcategories followed by *GRA&I* and *TAB*. For example, "Field 16, Group D, Missiles" subcategory in *GRA&I* is further divided in *TAB* into "1. Missiles, 2. Air-and-Space launched missiles, 3. Surface-launched missiles, and 4. Underwater-launched missiles.

Each issue of *TAB Indexes* contains Corporate Author-Monitoring Agency Index, Subject Index, Title Index, Personal Author Index, Contract Index, Report Number Index, and Release Authority Index. The indexes are cumulated quarterly and annually. Each issue of *TAB* contains changes in classification and distribution limitations of individual reports.

Scientific and Technical Aerospace Reports (*STAR*) covers reports in the areas of "aeronautics and space research and development, supporting basic and applied research, and applications." Also covered are "aerospace aspects of earth resources, energy development, conservation, oceanography, environmental protection, urban transportation, and other topics of high national priority."

Selected Water Resources Abstracts, another source of interest, covers reports dealing with the "water related aspects of the life, physical, and social sciences as well as related engineering and legal aspects of the characteristics, conservation, control, use, or management of water."

Energy Research Abstracts (*ERA*) includes "DOE's [i.e., Department of Energy's] research, development, demonstration and technological programs resulting from its broad charter for energy systems, conservation, safety, environmental protection, physical research and biology and medicine."

In contrast to *GRA&I* and *TAB*, the coverages of *STAR, Selected Water Resources Abstracts*, and *ERA* are not limited to technical reports alone. For example, *STAR* lists dissertations that are selected from *Dissertations Abstracts International; Selected Water Resources Abstracts* and *Energy Research Abstracts* both include coverage of journal articles, monographs, and conference reports.

AGENCY BIBLIOGRAPHIES

In addition to the comprehensive and mission-based access tools for technical reports, many agencies publish bibliographies covering their own technical reports. For example, *Bibliography of Technical Publications and Papers* (AD-A048 232) lists the reports "presented by personnel of the Army NATIC Research and Development Command and its contractors." Similarly, *Annotated Bibliography of Reports* (AD-A053 401) lists reports published by the Naval Aerospace Medical Research Laboratory (NAMRL). National Science Foundation announces technical reports resulting from its research programs in *Recent Research Reports* (e.g., PB80-201866, April 1980 and PB81-111767, April 1981).

Another type of agency publication is the bibliography issued by private organizations which may also include federally sponsored technical reports. An example of such a publication is *Selected Rand Abstracts* which includes reports issued by Rand Corporation and were sponsored by various federal agencies. These reports, in addition to being available from Rand Corporation, may also be available from National Technical Information Service (NTIS), Defense Technical Information Center (DTIC), National Aeronautics and Space Administration (NASA), and Educational Resources Information Center (ERIC) Document Reproduction Service.

INSTITUTIONAL SOURCES

Major institutional sources for technical reports include National Technical Information Service (NTIS) (5285 Port Royal Road, Springfield, VA 22161), Defense Technical Information Center (DTIC), formerly Defense Documentation Center (Cameron Station, Alexandria, VA 22314), NASA Scientific and Technical Information Facility (P.O. Box 8757, B.W.I. Airport, MD 21240), Technical Information Center, Department of Energy (P.O. Box 62, Oak Ridge, TN 37830), and Government Printing Office (Washington, DC 20402). These organizations provide access to technical reports by developing and promoting a variety of bibliographic products and services. A brief description of each of these facilities follows.

NTIS maintains a collection of more than one million unclassified technical reports. General public and researchers, irrespective of their affiliation, can purchase government-sponsored research and development reports from this organization. Also available from NTIS are machine-readable data files prepared by federal agencies and their grantees and contractors. In its efforts to promote the sale of the reports, NTIS provides the following services and products:

- Publishes *Government Reports Announcements & Index*
- Publishes the following 26 current awareness weekly abstract newsletters:
  Administration & Management
  Agriculture & Food
  Behavior & Society
  Biomedical Technology & Human Factors Engineering
  Building Industry Technology
  Business & Economics
  Chemistry
  Civil Engineering
  Communication
  Computers, Control & Information Theory
  Electrotechnology
  Energy
  Environmental Pollution & Control
  Government Inventions for Licensing
  Health Planning & Health Services Research
  Industrial & Mechanical Engineering
  Library & Information Sciences
  Materials Sciences
  Medicine & Biology
  NASA Earth Resources Survey Program
  Natural Resources & Earth Sciences
  Ocean Technology & Engineering
  Physics
  Problem-Solving Information for State & Local Governments
  Transportation
  Urban & Regional Technology & Development
- Maintains NTIS bibliographic data base which can be accessed online from DIALOG Information Retrieval Services (3460 Hillview Avenue, Palo Alto, CA 94303), ORBIT Information Retrieval System (2500 Colorado Avenue, Santa Monica, CA 90406), and Bibliographic Retrieval Services, Inc. (1200 Route 7, Latham, NY 12110). NTIS data base can also be leased directly from NTIS.
- Makes available bibliographic data files developed by other federal agencies, e.g., Energy Data Base (Department of Energy), Patent Data File (Patent and Trademark Office), Selected Water Resources Abstracts (Department of the Interior), and AGRICOLA (Department of Agriculture)

- Makes available machine-readable nonbibliographic datafiles and software (e.g., *National Drug Code Directory* and *Registry of Toxic Effects of Chemical Substances*)
- Makes available more than 3,500 published searches
- Acts as a marketing coordinator for a number of information analysis centers
- Acts as a subscription source for a number of printed abstracting services (e.g., *Fusion Energy Update, Oncology Overviews*, and *Selected Water Resources Abstracts*)
- Has a service called Selected Research in Microfiche (SRIM) which allows NTIS customers to purchase microfiche copies of reports by subject categories and subcategories
- Has a variety of sales services such as deposit accounts, rush handling, and online ordering
- Publishes a number of brochures explaining its products and services, e.g., *General Catalog of Information Services* and *Data Base Services and Federal Technology in Machine-Readable Formats* (C51.11/4:(date))

DTIC serves the Department of Defense and its contractors as well a other U.S. governmental agencies and their contractors. While NTIS deals only with unclassified/unlimited distribution reports, DTIC handles classified and limited distribution reports as well. Consequently, access to DTIC services is restricted and it requires prior registration with DTIC and certification of "need-to-know." The services and products of DTIC are as follows:

- Maintains a collection of more than 1.25 million reports in the subject areas such as aeronautics and missile technology as well as in basic sciences such as biology and chemistry
- Maintains four data bases which are: 1) the Research and Development Program Planning Data Base (R&DPP), 2) the Research and Technology Work Unit Information System (WUIS), 3) the Technical Reports Data Base (TR), and 4) the Independent Research and Development Data Base (IR&D). These data bases allow access to technical information at various stages of completion.
- Publishes *Technical Abstract Bulletin* (*TAB*) and *TAB* Indexes and a periodical called *DTIC Digest*
- Other services include: Automatic Document Distribution (ADD) Program, current awareness service program, preparation of on-demand bibliographies, reference and referral services
- Online access to the four DTIC data bases which is known as DROLS, Defense RDT&E On-Line System
- For further information: *User's Guide to: Defense Technical Information Center: Programs, Products, Services* (Washington, DC: Government Printing Office, 1980)

NASA Scientific and Technical Information Facility (STIF) maintains an extensive collection of reports and other types of documents in aeronautics and space sciences. The facility's primary clientele are NASA scientists and engineers. Just as in the case of DTIC, user organizations register with NASA STIF for service. The following are the information products and services of STIF:

- Maintains a collection of more than one million documents which includes journal articles, books, conference proceedings, translations, dissertations, NASA and non-NASA technical reports. These documents are collected and processed by STIF, American Institute of Aeronautics and Astronautics (AIAA), and European Space Agency (ESA).
- Publishes *Scientific and Technical Aerospace Reports* which announces unclassified reports and *Limited Scientific and Technical Aerospace Reports* (*LSTAR*) which announces classified and limited-distribution reports. Sponsors the publication of *International Aerospace Abstracts* (*IAA*) which complements *STAR* and *LSTAR* in that *IAA* indexes journal articles, conference papers, and other forms of published literature. Also published, until recently, *Computer Program Abstracts* (NAS1.44:(v.nos.&nos.))
- Issues current awareness service publications called Selected Current Awareness Notices (SCAN) in about two hundred subject categories. Documents announced in SCAN are obtained from *STAR* and *IAA*.
- Publishes a series of continuing bibliographies: *Aeronautical Engineering* (NAS1.21:7037(nos.)), *Aerospace Medicine and Biology* (NAS1.21:7011(nos.)), *Earth Resources* (NAS1.21:7041 (nos.)), *Energy* (NAS1.21:7043(nos.)), and *NASA Patent Abstracts* (NAS1.21:7039(nos.)). These are available to the general public.
- Maintains and provides online access to RECON (REmote CONsole) Data Base. For those organizations that do not have access to RECON, the Facility conducts individual literature searches on demand.
- Has Automatic Document Distribution Service (ADDS) which automatically distributes microfiche and paper copies of reports in eleven subject areas: aeronautics, astronautics, chemistry and materials, engineering, geosciences, life sciences, mathematical and computer sciences, physics, social sciences, space sciences, and general.
- NASA maintains eleven depository libraries which receive NASA technical publications. These are
  1. University of California, Berkeley.
  2. Carnegie Library of Pittsburgh.
  3. University of Colorado, Boulder.
  4. Columbia University, New York.
  5. Library of Congress.
  6. Georgia Institute of Technology, Atlanta.

7. The John Crerar Library, Chicago.
8. Linda Hall Library, Kansas City.
9. Massachusetts Institute of Technology, Cambridge.
10. University of Oklahoma, Bizzell Library, Norman.
11. University of Washington, Seattle.

- Detailed information on the facility's services can be found in *The NASA Information System: And How to Use It* (Washington, DC: National Aeronautical and Space Administration, The Scientific and Technical Information Branch, n.d.).

The DOE Technical Information Center (TIC) collects and indexes worldwide scientific and technical literature in the area of energy. It also collects and archives DOE-generated energy information and distributes these reports within DOE and to its contractors. Its services are

- Maintains the Energy Data Base (EDB) which contains nearly a million citations to worldwide energy literature. EDB is available online to the general public through BRS, DIALOG, and ORBIT systems. It is also available online through DOE/RECON system to DOE and DOE contractors. DOE/RECON allows access to Nuclear Science Abstracts (NSA) Data Base, Research in Progress, Electric Power Research Data Base (EPD), Federal Research Progress (FRP), Water Resources Research (WRE), Issues and Policy Summaries (IPS), Energy Information Resources Inventory (EIRI), Federal Energy Data Index (FEDEX), and other data bases with controlled access called Classified Data Base and Limited Reports Data Base.

- Issues the following publications:
  Energy Research Abstracts (*ERA*) (E1.17:(v.nos.&nos.)),
  Energy Abstracts for Policy Analysis (*EAPA*) (E1.11:(v.nos.& nos.)),
  Current Energy Patents (*CEP*),
  DOE Patents Available for Licensing (*PAL*),
  Energy and the Environment (*EAE*),
  Fossil Energy Update (*FEU*),
  Fusion Energy Update (*CFU*),
  Geothermal Energy Update (*GEU*),
  Solar Energy Update (*SEU*),
  Direct Energy Conversion (*DEC*),
  Nuclear Fuel Cycle (*NFC*),
  Nuclear Reactor Safety (*NRS*),
  and *Radioactive Waste Management* (*RWM*).
  The first two publications are available from GPO; the others are available from NTIS. The TIC cooperates with International Atomic Energy Agency, International Nuclear Information System (IAEA/INIS) in publishing *Atomindex*. International Energy Agency, IEA Coal Research publishes *Coal Abstracts* in cooperation with TIC. Other TIC publications are *Abstracts of Weapon Data Reports*, which is based on the Center's Classified Data Base, and *Energy Meetings*.

- Automatically distributes appropriate reports to DOE contractors. Unclassified distribution reports are sold through NTIS and a TIC microforms contractor: Engineered Systems, Inc., P.O. Box 866, Oak Ridge, TN 37830. Distributes to the public, public information documents published by DOE and its contractors.
- Makes available DOE and DOE contractor-produced computer software packages through National Energy Software Center under the direction of TIC.
- Detailed information on TIC services can be found in *Technical Information Center: Its Functions and Services* (Oak Ridge, TN: U.S. Department of Energy, Technical Information Center, May 1982) (DOE/TIC-4600) (Rev.3). Further, a number of libraries maintain extensive collections of DOE reports. These libraries are listed in the issues of *Energy Research Abstracts.*

The Superintendent of Documents, Government Printing Office (GPO) announces numerous technical reports through *Monthly Catalog* whether they are for sale from GPO or not. One can locate publications currently on sale from GPO through GPO Sales Publications Reference File (PRF), a microfiche catalog. The Publications Reference File is accessible on DIALOG and users can place orders online through this system. *New Books* (bimonthly) (GP3.17/6:(v.nos.&nos.)) is a new periodical issued by the GPO, and it lists the documents placed on sale during the preceding two-month period.

Another source of information from GPO is *U.S. Government Books* (GP3.17/5:(v.nos.&nos.)), which lists publications of popular interest. This periodical supersedes *Selected U.S. Government Publications* (GP3.17:(v.nos.&nos.)). Popular titles of GPO publications are also sold through twenty-six GPO bookstores located in various cities. In any case, a significant number of technical reports are made available to the depository libraries through the depository library program of the Government Printing Office.

DUPLICATION AMONG MAJOR TECHNICAL REPORTS' BIBLIOGRAPHIC SOURCES

There is a significant amount of overlap among the major bibliographic sources for technical reports, namely *Government Reports Announcements & Index, Scientific and Technical Aerospace Reports, Energy Research Abstracts,* and *Monthly Catalog.* Copeland, in a recent study, concluded that *GRA&I* indexed 94.3 percent of technical reports that appeared in *STAR* and 78.8 percent of those that appeared in *ERA.*[24]

Copeland further concluded that about 41 percent and 21 percent of technical report entries of *GRA&I* appeared in *STAR* and *ERA* respectively. This indicates, not surprisingly, that *GRA&I* is the major source for information on technical reports. This, of course, is a useful observation for reference librarians in that the print or online versions of *GRA&I* should be the first source to consult for locating information on technical reports followed by other indexes.

There is also some overlap between *Monthly Catalog* and *GRA&I.* This overlap is great from the perspective of *Monthly Catalog*, although it is not as great from the perspective of *GRA&I.* Nevertheless, this overlap between these two sources has implications for reference service.

The fact that the technical reports are products of federal involvement in scientific research and development brings them under the purview of documents collections while their scientific content is of interest to science and technology libraries. In this connection, two factors cause confusion: 1) technical reports, being federally produced or sponsored publications, fall into the scope of both *GRA&I* and *Monthly Catalog*, and 2) the location mechanism of *Monthly Catalog*-based documents collections differ from that of *GRA&I*-based technical report collections. These two factors create the illusion that technical reports are distinct from government documents.

The fact that technical reports *are* government documents is reflected in the extent of duplication between *Monthly Catalog* and *GRA&I*. An examination of a recent issue of the *Monthly Catalog* shows that 30 percent of the items listed in it are available from NTIS. This number, however, under-represents both the extent of *Monthly Catalog-GRA&I* duplication and the number of technical reports included in the *Monthly Catalog*. For example, a number of publications from agencies such as National Bureau of Standards are routinely available from NTIS, but this fact is not always indicated in the *Monthly Catalog*. Similarly, if one considers the *Monthly Catalog* entries under descriptors such as technical note, research note, and research paper, as well as the titles of documents for scientific and technical content, one can estimate that as high as 40 percent of the items listed in *Monthly Catalog* are technical reports.

However, this duplication between *Monthly Catalog* and *GRA&I* is not always obvious. For example, *Monthly Catalog* does not mention that a document whose report number is NBS-SP-596 is available both in paper copy and microfiche formats from NTIS; at the same time, *GRA&I* does not mention that it is available for sale from the Superintendent of Documents. Further, in a federal documents collection, NBS-SP-596 is physically located under the Superintendent of Documents classification number, C13.10:596. The same report is given an accession number PB81-124984 by NTIS. Neither *Monthly Catalog* nor *GRA&I* attempts to link these two location symbols. (See figures 4.1 and 4.2, page 54, for duplications).

This duplication between *Monthly Catalog* and *GRA&I* is not necessarily undesirable. In fact, such a duplication provides an improved access to the technical reports—especially if one is aware of the extent of duplication.

Some of the reports announced in the *Monthly Catalog* are depository items and are sent to the GPO depository libraries that selected these items. In other words, these technical reports, under the Superintendent of Documents classification numbers, are widely available across the country. Examples of such technical reports are shown in figure 4.3 (page 54). Similarly, many of the technical reports announced in the *Monthly Catalog* are also available from commercial sources such as Readex Microprint Corporation. The duplication between the *Monthly Catalog* and *GRA&I* assures the availability of technical reports long after they are out of print at GPO.

Finally, the availability of *Monthly Catalog* and *GRA&I* for online searching is likely to spur the greater use of GPO documents collections by the traditional users of NTIS report collections and vice versa. Of course, this type of cross-use could be improved greatly by comprehensive cross referencing between Superintendent of Documents classification numbers, NTIS accession/report numbers, as well as between the entries of *Monthly Catalog* and *GRA&I*.

FIGURE 4.1
MONTHLY CATALOG AND *GRA&I* ENTRIES: A COMPARISON

81-4277
C13.10:596
International Symposium on Ultrasonic Materials Characterization (1st : 1978 : National Bureau of Standards)
Ultrasonic materials characterization : proceedings of the first International Symposium on Ultrasonic Materials Characterization held at the National Bureau of Standards, Gaithersburg, Md., June 7-9, 1978 / co-editors: Harold Berger, Melvin Linzer.-[Washington, D.C.?] : U.S. Dept. of Commerce, National Bureau of Standards : For sale by the Supt. of Docs., U.S. G.P.O., 1980.
xi, 663p. : ill. ; 27 cm. - (NBS special publication ; 596)
"National Measurement Laboratory, National Bureau of Standards."
"Issued November 1980."
Includes bibliographical references.
• Item 247
S/N 003-003-02264-1 @ GPO
$11.00
1. Ultrasonics - Congresses. I. Berger, Harold. II. Linzer, Melvin. III. National Measurement Laboratory. IV. Title. V. Series.
80-600148
OCLC 7185465

PB81-124984 PC A99/MF A01
National Bureau of Standards, Washington, DC.
ULTRASONIC MATERIALS CHARACTERIZATION
Final rept.,
Harold Berger, and Melvin Linzer. Nov 80, 649p
NBS/SP-596
Proceedings of the First International Symposium on Ultrasonic Materials Characterization held at the National Bureau of Standards, Gaithersburg, MD on June 7-9, 1978. Library of Congress catalog card no. 80-600148.

Nondestructive testing has traditionally involved a search for flaws in materials or structures; it has long been appreciated that voids, cracks, inclusions, and similar flaws can lead to failure....

FIGURE 4.2
AN EXAMPLE OF A DOCUMENT WHICH IS BOTH
A DEPOSITORY ITEM AND AN NTIS REPORT

81-10640

E1.19:0092

Capsule review of the DOE research and development and field facilities. - Washington, D.C. : U.S. Dept. of Energy, Office of Energy Research, Office of Field Operations Management ; Springfield, Va. (5285 Port Royal Rd., Springfield, Va., 22161) : Available from National Technical Information Service, 1981.

vi, 43 p. : ill., maps ; 28 cm.

Sept. 1980

"DOE/ER-0092."

• Item 429-N

1. Power resources - Research - United States. 2. Energy conservation-Research-United States. I. United States. Dept, of Energy. Office of Energy Research. Office of Field Operations Management.

OCLC 07632090

FIGURE 4.3
NATIONAL AERONAUTICS AND SPACE ADMINISTRATION'S
TECHNICAL REPORTS AVAILABLE TO DEPOSITORY LIBRARIES*

Series Name	*Item Number*	*SupDoc No*
Jet Propulsion Laboratory: JPL/HR-	0830-H-03	NAS 1.12/4:
Jet Propulsion Laboratory: SP	0830-H-03	NAS 1.12/6:
Jet Propulsion Laboratory: Mission Status Bulletins	0830-Y	NAS 1.12/8:
Technical Memorandums (MF)	0830-D	NAS 1.15:
Handbooks, Manuals, Guides (P)	0830-F	NAS 1.18:
NASA EP (series)	0830-G	NAS 1.19:
NASA SP (series) (with exceptions) (P)	0830-I	NAS 1.21:
Reliability and Quality Assurance Publications (NHB series) (MF)	0830-M	NAS 1.22/3:
NASA Contractor Reports (numbered) (MF)	0830-H-14	NAS 1.26:
Mission Reports, MR-(series)	0830-Y	NAS 1.45:
Conference Publications NASA CP (series)	0830-H-10	NAS 1.55:
NASA Technical Papers (numbered) (MF)	0830-H-15	NAS 1.60:

*Source: U.S. Government Printing Office. Depository Administration Branch, *List of Classes of United States Government Publications Available for Selection by Depository Libraries* (Washington, DC: GPO, 1982) (GP3.24:982/3).

REFERENCES

1. For information, see Eugene B. Jackson, *Unpublished Research Reports: A Problem in Bibliographical Control* (University of Illinois Library School Occasional Papers, No. 17) (Urbana, IL: University of Illinois Library School, December 1950) and Johanna Tallman, "History and Importance of Technical Reports," *Sci-Tech News* 15 (Summer 1961): 44-46; 15 (Winter 1962): 164-65, 168-72; and 16 (Spring 1962), 13.
2. U.S. President's Science Advisory Committee, *Science, Government, and Information* (Weinberg Report) (Washington, DC: Government Printing Office, 1963), p. 19.
3. H. Skolnik, "References Are Not Equal" (Editorial), *Journal of Chemical Documentation* 10 (May 1970): 74.
4. King Research Inc., *A Chart Book of Indicators of Scientific & Technical Communication in the United States* (Prepared for: Division of Information Science and Technology, National Science Foundation) (Washington, DC: Government Printing Office, 1978), p. 7.
5. U.S. Department of Defense, Defense Documentation Center, *Glossary of Information Handling* (Arlington, VA, 1964).
6. U.S. National Science Foundation, *Federal Funds for Research and Development: Fiscal Years 1980, 1981, and 1982* (NSF 81-325) (Washington, DC: Government Printing Office, 1981), p. 14.
7. U.S. National Science Foundation, *An Analysis of Federal R&D Funding by Function: Fiscal Years 1969-1979* (Washington, DC: Government Printing Office, 1979), pp. 6, 34-35.
8. Neil Bearley, "The Role of Technical Reports in Scientific and Technical Communication," *IEEE Transactions on Professional Communication* PC-16 (September 1973): 117-19.
9. Ibid., p. 117.
10. Dwight E. Gray and Steffan Rosenborg, "Do Technical Reports Become Published Papers?" *Physics Today* 10 (June 1957): 18-21.
11. Ibid., p. 18.
12. William D. Garvey, *Communication: The Essence of Science* (Oxford, England: Pergamon, 1979), p. 59.
13. For information, see Gray and Rosenborg, op. cit., p. 19 and Garvey, op. cit., p. 60.
14. U.S. Committee on Scientific and Technical Information, "Task Group on the Role of the Technical Report," in *The Role of the Technical Report in Scientific and Technological Communication* PB 180 944 (Washington, DC, December 1968), p. 26.
15. Garvey, op. cit., p. 60.

16. W. D. Garvey and B. C. Griffith, "Communication and Information Processing within Scientific Disciplines: Empirical Findings for Psychology," in Garvey, op. cit., pp. 127-47.

17. Gray and Rosenborg, op. cit., p. 21.

18. For information, see Gray and Rosenborg, op. cit., p. 21 and COSATI, "Task Group on the Role of the Technical Report," op. cit., p. 26.

19. W. D. Garvey, Nan Lin, and K. Tomita, "Research Studies in Patterns of Scientific Communication: III. Information-Exchange Process Associated with the Production of Journal Articles," in Garvey, op. cit., pp. 202-24.

20. Garvey, op. cit., p. 59.

21. COSATI, "Task Group on the Role of the Technical Report," op. cit., pp. 28-30.

22. C. W. J. Wilson, "Obsolescence of Report Literature," *Aslib Proceedings* 16 (June 1964): 200-1.

23. H. Voos, "Information Explosion: or, Redundancy Reduces the Charge!" *College and Research Libraries* 32 (January 1971): 7-14.

24. Susan Copeland, "Three Technical Report Printed Indexes: A Comparative Study," *Science & Technology Libraries* 1 (Summer 1981): 41-53.

5

PERIODICALS

As a means of communicating scientific and technical information, federal agencies publish or sponsor several newsletters, magazines and research journals in a variety of subject areas such as health sciences, engineering, and agriculture. The periodicals published by the federal agencies vary widely in terms of intended audience, technical level of the subject matter, type of publication, and agencies responsible for publication. The intended audience for the federal scientific periodicals includes scientists, both inside and outside the government, federal employees such as technicians, and the general public.

The contents of these publications vary from highly technical to popular level. Types of publications include primary journals that report results of original research, reviews of research, newsletters, and official magazines that inform the public of agencies' policies and programs. Finally, the organizations responsible for publication of these periodicals include federal agencies, federally funded information analysis centers, and research laboratories.

A number of federal periodicals are aimed at practicing scientists. *Environmental Health Perspectives* and the *Journal of Research of the National Bureau of Standards* are the representatives of the primary journals that publish original research observations. These journals follow the typical scientific journal format and policies such as subjecting the articles to peer review before publication. They report on research which either has been conducted in federal laboratories or has been funded by federal agencies.

Primary journals, however, form only a small part of federal scientific and technical periodicals. This is because of the tendency of the scientists and engineers, whether they are federal employees or those supported by federal grants and contracts, to publish in standard scientific journals in their fields most of which are published by non-federal sources such as nonprofit scientific associations and commercial publishers. For example, scientists and engineers at the National Bureau of Standards recently published in non-federal periodicals such as

American Chemical Society. Polymer Preprints
American Minerologist
Analytical Chemistry
Analytica Chimica Acta
Astrophysical Journal
Biopolymers
Chemical Physics
Chromatographica
Geochemical Journal
International Journal of Mass Spectrometry and Ion Physics
Journal of the Acoustical Society of America
Journal of the American Chemical Society
Journal of Applied Physics
Journal of the Association of Official Analytical Chemists
Journal of Chemical Physics
Journal of Chromatography
Journal of Physics B: Atomic and Molecular Physics
Molecular Physics
Neurology
Physical Review A
Science

Professional newsletters are another type of periodical addressed to working scientists. *NSRDS Reference Data Report* is an example of such a newsletter. Typically, such newsletters contain brief reports on research in progress, announcements of publications and forthcoming conferences, and information on fellow scientists.

A number of other periodicals containing technical information are intended specifically for the respective agency employees. These periodicals attempt to serve one of two functions: 1) keeping the employees informed of personnel changes and work taking place in the organization and 2) communicating new developments and techniques in their respective technical areas. *ARRADCOM Voice* serves the first purpose; *Approach: The Naval Aviation Safety Review* serves the second purpose.

Finally, periodicals such as *EPA Journal* are designed to inform the general public of the agencies' policies and programs. In addition to general-interest articles, these magazines contain interviews and news items on their respective agencies. Usually, these are issued by departments such as the Office of Public Awareness and the Office of Public Information.

Many of the journals mentioned above are published by federal agencies or their subdivisions. Many other periodicals are issued by organizations outside the government which are nevertheless supported by federal monies. Examples of such outside organizations include the federal grantees and contractors, federally supported information analysis centers, and research and developmental laboratories. In many cases, such non-governmental organizations are operated by universities and other nonprofit organizations or by industrial organizations. However, these agencies form an integral part of the scientific and technical information generated by the federal government.

An example of such an organization is the Applied Physics Laboratory, operated by the Johns Hopkins University. The laboratory is supported by funds from the Navy and other federal departments. One of its affiliates, the Chemical Propulsion Information Agency (CPIA), publishes a bimonthly called *Chemical Propulsion Newsletter*.

The Journal of Physical and Chemical Reference Data is an example of a periodical published by scientific associations for a government agency. It is published by the American Chemical Society and the American Institute of Physics for the National Bureau of Standards. The journal, which publishes "critically evaluated physical and chemical property data," is one of the principal publication media for the National Bureau of Standards.

Finally, some journals that were once published by the federal government, are currently published by non-governmental agencies. These include *Monthly Weather Review* (published by the American Meteorological Society) and *Radio Science* (published by the American Geophysical Union). The *Monthly Weather Review* was published by various federal agencies for 101 years before it was turned over to the American Meteorological Society because, during the last years of federal sponsorship, its contributors had represented a "cross-section of the entire meteorological community, both national and international." *Radio Science* had its origins in Section D of the *Journal of Research of the National Bureau of Standards*. From 1959 through 1968, it was a federal publication. Even to this day, the principal source of articles in these journals continues to be federally funded research projects, and articles from federal scientists.

NON-FEDERAL PERIODICALS

Federally published scientific periodicals are, in fact, only minor means of transmitting results of federally produced or sponsored research. Non-federal scientific journals play a vital role in communicating scientific information, a significant part of which is produced with federal support. Many of the prestigious scientific journals, published either by nonprofit organizations or by commercial publishers, depend heavily on federally sponsored research.

A cursory examination of well-known scientific journals such as the *Journal of the American Chemical Society, Journal of Biological Chemistry, Proceedings of the National Academy of Sciences*, and *Physical Review* bears out the extent of their dependence on the federally sponsored research. A large portion of the research published in these journals is supported by the agencies such as National Science Foundation, Department of Energy, National Institutes of Health, and the Public Health Service.

As was mentioned previously, agencies themselves publish their research in non-federal publications. For example, in 1981 nearly half of the 1,767 papers published by the National Bureau of Standards appeared in non-NBS sources, most of which are non-federal journals. Similarly, one can also notice the symbiotic relationship between federal agencies and the periodicals published by the commercial and nonprofit organizations by the fact that federal scientists are on the editorial boards of these periodicals. For example, the editor-in-chief and an associate editor of the *Journal of Biological Response Modifiers* are federal scientists. They work for the National Cancer Institute's Biological Response Modifiers Program.

The vital role such non-federal journals play in furthering scientific communication can be appreciated by the fact that universities and colleges account for half of all the expenditures in basic research in the United States and most of these funds come from the federal government. Much of the information generated by this basic research finds its way into the non-federal scientific journals.

INDEXING AND ABSTRACTING SERVICES COVERING FEDERAL PERIODICALS

The only service that is devoted entirely to indexing government periodicals is the *Index to U.S. Government Periodicals: A Computer Generated Guide to Selected Titles by Author and Subject*. Unfortunately, this index dates back only to 1974. Discipline-based indexing services such as *Chemical Abstracts, Biological Abstracts, Engineering Index, Index Medicus*, and *Science Citation Index* cover substantive federal science periodicals in their respective areas of interest.

Other bibliographic sources for articles published by federal scientists in non-federal publications include agency bibliographies such as *Publications of the National Bureau of Standards* (Special Publication 305 and its supplements) (C13.10:305/(supp.)) and the *NBS Publications Newsletter* (C13.36/5, monthly); *National Institutes of Health: Scientific Directory: Annual Bibliography* (HE20.3017:(yr.)); and *NCI Grant Supported Literature Index* (HE20.3179:(date)).

ACQUISITION AND OTHER INFORMATION SOURCES

In the course of reference or documents work, a librarian often needs a variety of bibliographic and subscription information on federal periodicals. *Government Periodicals and Subscription Services*, also called Price List 36 (GP3.9:36(nos.)), is a source of current information on periodicals, printed by GPO and available for sale from the Superintendent of Documents.

Another source of subscription information is the Serials Supplement of the *Monthly Catalog of United States Government Publications* (GP3.8:(yr./supp.)). The recent supplements contain full cataloging information for publications issued more than three times per year, as well as a select group of annual publications. The supplement has author, title, and subject indexes to provide access to the main section which is arranged by the Superintendent of Documents number. The supplement, however, cannot be depended upon completely in view of the fact that a number of periodicals listed are likely to have been discontinued.

Current availability of periodicals from GPO can be ascertained by using *GPO Sales Publications Reference File* (GP3.22/3). The reference file is used bimonthly in microfilm form. Periodicals can be located in this file by means of GPO stock numbers, Superintendent of Documents classification numbers, titles, and under the headings "Government Periodicals and Subscription Services."

For retrospective information on these periodicals, one of the most important sources is the *Guide to U.S. Government Publications* (John L. Andriot, ed. McLean, VA: Documents Index, 1962-). It is arranged by the Superintendent of Documents classification number. If one is interested in following the changes a periodical has undergone, in terms of publishing agency, frequency of publication, etc., the guide is a very useful source.

Another useful source for retrospective information is *U.S. Government Serial Titles, 1789-1976: Index IV to the Dual Media Edition of Checklist of United States Public Documents, 1789-1976* (Arlington, VA: United States Historical Documents Institute, 1978) and its supplements. *Serial Titles*, which is an annotated alphabetical listing of current and discontinued titles in the Serials Card File of the U.S. Superintendent of Documents' Public Documents Library, also helps in tracing the histories of the federal periodicals, including title and class number changes.

None of the sources mentioned above, however, is helpful in locating information about the periodicals issued by the federal grantees, contractors, and federally supported organizations. To a limited extent, *Government Research Centers Directory*, 2d ed. (Detroit: Gale Research Company, 1982) is useful in this regard. Although the directory does not provide bibliographic information, it does list the periodicals issued by each center.

Similarly, *The Encyclopedia of Information Systems and Services*, 5th ed. (Detroit: Gale Research Company, 1982) may also be useful in locating such periodicals. Generally, most of these periodicals can be obtained by writing directly to the agency involved. Finally, general sources such as *New Serial Titles* (Washington, DC: Library of Congress, 1953-) and *Ulrich's International Periodicals Directory* (New York: Bowker, 1932-) do list a number of federal publications in them.

The following are examples of scientific periodicals published by the federal sources. They are listed under four categories: 1) primary research journals, 2) professional newsletters, 3) employee training/education periodicals, and 4) public awareness magazines.

RESEARCH JOURNALS

Title: Air Force Journal of Logistics, v. 1, no. 1, 1977- . D301.91:(v.nos.&nos.)
Producer: Air Force Logistics Management Center, Gunter AFS, AL 36114
Scope: Publishes articles and research results dealing with various aspects of Air Force logistics such as maintenance, supply, transportation, and logistics plans. Has a regular section called Current Research which lists research projects in progress.
Frequency: Quarterly
Availability: Superintendent of Documents, depository item

Title: Bulletin of Prosthetics Research, no. 1, Spring 1964- . VA1.23/3.10-(nos.)

Producer: Rehabilitative Engineering Research and Development Service, Department of Medicine and Surgery, Veterans Administration, Washington, DC 20420. Editorial correspondence: Editor, *Bulletin of Prosthetics Research*, Office of Technology Transfer (153D), Veterans Administration, 252 Seventh Avenue, New York, NY 10001.

Scope: It is a refereed journal and is a source of information on research and development activities in rehabilitative engineering area. Also includes technical notes, progress reports, abstracts of recent articles, recent patents, publications of interest, calendar of events, conference reports, and notes and news.

Frequency: Semiannual

Availability: Superintendent of Documents

Title: Cancer Treatment Reports, v. 1, 1959- . HE20.3160:(v.nos.&nos.)

Producer: National Cancer Institute, 7910 Woodmont Avenue, Room 8C08, Landow Bldg., Bethesda, MD 20814

Scope: Publishes clinical and laboratory research articles on cancer treatment methods such as radiotherapy, chemotherapy, and immunotherapy.

Frequency: Monthly

Availability: Superintendent of Documents, depository item

Title: Clinch River Breeder Reactor Plant Technical Review, Spring 1976- .

Producer: Breeder Reactor Corporation, P.O. Box U, Oak Ridge, TN 37830

Scope: Reports on research being conducted on the Clinch River Breeder Reactor plant project, which is designed to demonstrate the feasibility of a large-scale liquid metal fast breeder reactor as a commercial source of electric power. Articles are fairly technical.

Frequency: Quarterly

Availability: Breeder Reactor Corporation, Oak Ridge, TN

Title: The Coast Guard Engineer's Digest, no. 1, January 1939- . TD5.17:(nos.)

Producer: Office of Engineering, U.S. Coast Guard Headquarters (CG-Eg/TP62), Washington, DC 20593

Scope: Informal in tone, the *Digest* publishes research papers as well as general-interest articles. Examples are: machinery plant of a new A class cutter, fuel cell technology report, service contracting, and all-weather piloting for mariners.

Frequency: Quarterly

Availability: Coast Guard Headquarters, depository item

Title: Concepts: The Journal of Defense Systems Acquisition Management, 1976- . D1.53:(v.nos.&nos.)

Producer: Defense Systems Management College, Fort Belvoir, VA 22060

Scope: Provides information on "policies, trends, events, and current thinking affecting program management and defense systems acquisition." Papers deal with procurement programs, weapons acquisition programs, costs of acquisitions and evaluations of equipment and services.
Frequency: Quarterly
Availability: Defense Systems Management College, depository item

Title: Environmental Health Perspectives, no. 1, April 1972- . HE20.3559(v.nos.)
Producer: National Institute of Environmental Health Sciences, National Institutes of Health, Public Health Service, Department of Health and Human Services, P.O. Box 12233, Research Triangle Park, NC 27709
Scope: It is a refereed journal and aims to inform the environmental science community of potential health hazards associated with various environmental agents. Includes conference and workshop proceedings as well as review articles on specific environmental problems and agents.
Frequency: Six volumes per year
Availability: Superintendent of Documents, depository item

Title: Fire Management Notes: An International Quarterly Periodical Devoted to Forest Fire Management, v. 1, no. 1, December 1936- . A13.32:(v.nos.&nos.)
Producer: Forest Service, U.S. Department of Agriculture, P.O. Bos 2417, Washington, DC 20013
Scope: Publishes articles and investigations of interest to forest fire-fighting personnel. Deals with topics such as fire prevention and control, fire safety, fire management, equipment, and training.
Frequency: Quarterly
Availability: Superintendent of Documents, depository item

Title: Fishery Bulletin, v. 1, 1881- . C55.313:(v.nos.&nos.)
Producer: Scientific Publications Office, National Marine Fisheries Service, National Oceanic and Atmosphere Administration, U.S. Department of Commerce, Washington, DC 20235
Scope: Carries original research reports in the areas of fishery science, engineering, and economics
Frequency: Quarterly
Availability: Superintendent of Documents, depository item

Title: Highway Focus, June 1969- . TD2.34:(v.nos.&nos.)
Producer: Construction and Maintenance Division, Federal Highway Administration, Department of Transportation, Washington, DC 20590
Scope: Publishes articles on highway engineering and construction operations having some unique features
Availability: Federal Highway Administration, depository item

Title: Journal of Physical and Chemical Reference Data, v. 1, 1972- .
Producer: American Chemical Society and American Institute of Physics for the National Bureau of Standards. ACS, 1155 Sixteenth Street, NW, Washington, DC 20056
Scope: It is the principal publication outlet for the National Standard Reference Data System (NSRDS). Publishes critically evaluated physical and chemical property data.
Frequency: Quarterly
Availability: American Chemical Society

Title: Journal of Research of the National Bureau of Standards, v. 1, no. 1, July 1928- . C13.22:(v.nos.&nos.)
Producer: National Bureau of Standards, U.S. Department of Commerce, Washington, DC 20234
Scope: Publishes research reports in the areas of physics, chemistry, engineering, mathematics and computer sciences.
Frequency: Bimonthly
Availability: Superintendent of Documents, depository item

Title: Journal of the National Cancer Institute, v. 1, no. 1, August 1940- . HE20.3161:(v.nos.&nos.)
Producer: National Cancer Institute, National Institutes of Health, Westwood Bldg., Room 850, 5333 Westbard Avenue, Bethesda, MD 20816
Scope: Publishes original research papers dealing with cancer clinical and laboratory research, epidemiology, and cancer control and prevention
Frequency: Monthly—two volumes per year
Availability: Superintendent of Documents, depository item

Title: Marine Fisheries Review, v. 1, no. 1, January 1939- . C55.310:(v.nos.&nos.)
Producer: Scientific Publications Office, National Marine Fisheries Service, NOAA, 7600 Sand Point Way NE, Bin C15700, Seattle, WA 98115
Scope: Publishes "review articles, original research reports, technical notes, and news articles on fisheries science, engineering, and economics, commercial and recreational fisheries, marine mammal studies, aquaculture, and U.S. and foreign fisheries developments." Emphasizes practical aspects of fisheries science.
Frequency: Monthly
Availability: Superintendent of Documents, depository item

Title: Nuclear Safety, v. 1, no. 1, September 1959- . E1.93:(v.nos.&nos.)
Producer: Prepared by: Nuclear Safety Information Center, P.O. Box Y, Oak Ridge National Laboratory, Oak Ridge, TN 37830. Published by: Technical Information Center, Department of Energy, Oak Ridge, TN 37830.

Scope: Covers topics "relevant to the analysis and control of hazards associated with nuclear energy, operations involving fissionable materials, and the products of nuclear fission, and their effects on the environment. Primary emphasis is on safety in reactor design, construction, and operation;... "
Frequency: Bimonthly
Availability: Superintendent of Documents, depository item

Title: The Progressive Fish-Culturist: A Quarterly for Fishery Biologists and Fish-Culturists, no. 1, December 1934- . I49.35:(v.nos.&nos.)
Producer: U.S. Fish and Wildlife Service, Aylesworth Hall, CSU, Fort Collins, CO 80523
Scope: A refereed journal dealing with fish and fish cultures
Frequency: Quarterly
Availability: Superintendent of Documents, depository item

Title: Psychopharmacology Bulletin, v. 1, no. 1, January 1959- . HE20.8109:(v.nos.&nos.)
Producer: Pharmacologic and Somatic Treatments Research Branch, National Institute of Mental Health, Department of Health and Human Services, 5600 Fishers Lane, Rockville, MD 20857
Scope: Publishes research and review articles as well as symposia papers. Also includes book reviews and a recurring bibliography on psychopharmacology.
Frequency: Quarterly
Availability: Superintendent of Documents, depository item

Title: Public Health Reports: Official Journal of the U.S. Public Health Service, v. 1, 1878- . HE20.6011:(v.nos.&nos.)
Producer: Public Health Service, U.S. Department of Health and Human Services, Room 814, Reporters Bldg., Washington, DC 20201
Scope: Publishes research articles dealing with all aspects of public health. Articles deal with measles, vaccinations, sexually transmitted diseases, asbestos, food-borne illnesses and such public health matters.
Frequency: Bimonthly
Availability: Superintendent of Documents, depository item

Title: Public Roads: A Journal of Highway Research and Development, v. 1, no. 1, May 1918- . TD2.19:(v.nos.&nos.)
Producer: Offices of Research and Development, Federal Highway Administration, Washington, DC 20590
Scope: Contains original research articles dealing with topics such as computerized highway traffic control, aerodynamic behavior of bridges and earthquake damage to highway systems. Also has regular features titled: Recent Research Reports, Implementation/User Items and New Research in Progress.
Frequency: Quarterly

Availability: Superintendent of Documents, depository item

Title: Research Bulletin, v. 1, no. 1, 1979- .
Producer: SEA/Cooperative Research Program, College of Agriculture, Southern University, P.O. Box 9614, Southern Branch Post Office, Baton Rouge, LA 70813
Scope: Includes reports of basic and applied agriculture research conducted at the College of Agriculture, Southern University
Frequency: Irregular
Availability: College of Agriculture, Southern University

Title: Schizophrenia Bulletin, no. 1, December 1969- . HE20.8115:(v.nos.&nos.)
Producer: Center for Studies of Schizophrenia, National Institute of Mental Health, Department of Health and Human Services, 5600 Fishers Lane, Rockville, MD 20857
Scope: Includes articles and reviews on schizophrenia research and treatment. Contains a section called Bibliography and Abstracts on Schizophrenia. In general, each issue focuses on one topic, e.g., paranoia.
Frequency: Quarterly
Availability: Superintendent of Documents, depository item

Title: The Shock and Vibration Digest, v. 1, No. 1, April 1969- . D4.12/2:(v.nos.&nos.)
Producer: The Shock and Vibration Information Center, Naval Research Laboratory, Code 5804, Washington, DC 20375
Scope: Publishes tutorial and literature reviews dealing with sound, shock, and vibration technology. Also contains abstracts of current literature, news, book reviews, reviews of conferences and lists of short courses planned.
Frequency: Monthly
Availability: Shock and Vibration Information Center

Title: U.S. Navy Medicine, v. 1, 1943- . D206.7:(v.nos.&nos.)
Producer: Navy Medical Command, U.S. Navy, Washington, DC 20372
Scope: Contains articles in the areas of medicine, dentistry, and allied health sciences
Frequency: Monthly
Availability: Superintendent of Documents, depository item

Title: Water Operation and Maintenance Bulletin, no. 1, November 1952- . I27.41:(nos.)
Producer: Division of Operation and Maintenance Technical Services, Engineering and Research Center, Bureau of Reclamation, Denver, CO 80225
Scope: Provides operation and maintenance information to those operating water supply systems
Frequency: Quarterly
Availability: Bureau of Reclamation

PROFESSIONAL NEWSLETTERS

Title: The AFIP Letter, v. 1, 1951- .
Producer: Armed Forces Institute of Pathology, Department of the Army, Washington, DC 20306
Scope: Reports abstracts of recent papers in the area of pathology and provides information on continuing education courses offered by the institute
Frequency: Monthly
Availability: Armed Forces Institute of Pathology

Title: Alcohol, Health and Research World, v. 1, no. 1, Spring 1973- . HE20.8309:(v.nos.&nos.)
Producer: National Institute on Alcohol Abuse and Alcoholism, Public Health Service, P.O. Box 2345, Rockville, MD 20852
Scope: Designed for "counselors, researchers, educators, health professionals and others concerned with alcohol use and abuse." Provides practical information.
Frequency: Quarterly
Availability: Superintendent of Documents, depository item

Title: Antarctic Journal of the United States, v. 1, no. 1, January-February 1966- . NS1.26:(v.nos.&nos.)
Producer: National Science Foundation, Washington, DC 20550
Scope: Publishes a variety of news items of interest to Antarctic researchers. Also includes announcements of publications as well as research grants awarded by the foundation.
Frequency: Quarterly with a fifth annual review issue
Availability: Superintendent of Documents, depository item

Title: ASIAC Newsletter, v. 1, no. 1, March 1974- .
Producer: Aerospace Structures Information and Analysis Center, Wright-Patterson AFB, OH 45433
Scope: Contains brief articles of interest to the users of ASIAC. For example, recent issues described the major data bases searched by ASIAC as well as the concept of literature searching. Also publishes features such as Calendar of Events and Call for Papers.
Frequency: Irregular
Availability: ASIAC

Title: BCTIC Newsletter, no. 1, October 1975- .
Producer: Biomedical Computing Technology Information Center, R-1302, Vanderbilt Medical Center, Nashville, TN 37232
Scope: Includes brief news items and announces new publications and computer codes as well as upcoming meetings and courses
Frequency: Irregular
Availability: BCTIC

Title: BRH Bulletin, v. 1, 1967- . HE20.4103:(v.nos.&nos.)
Producer: Bureau of Radiological Health, Food and Drug Administration, Rockville, MD 20857
Scope: Reports on bureau activities in the form of brief news items. Also includes Calendar of Events and New Bureau Publications.
Frequency: Monthly
Availability: Bureau of Radiological Health, FDA, depository item

Title: Center for Water Quality Modeling Newsletter, 1980- .
Producer: Environmental Research Laboratory, Environmental Protection Agency, Athens, GA 30613
Scope: Publishes news items on water quality modeling. Announces forthcoming meetings and workshops as well as news items on modeling software and documentation.
Frequency: Quarterly
Availability: Environmental Research Laboratory

Title: Comparative Pathology Bulletin, v. 1, November 1969- .
Producer: Registry of Comparative Pathology, Armed Forces Institute of Pathology, Washington, DC 20306
Scope: It is a 4-page newsletter containing brief professional articles such as albino animals: their use and misuse in biomedical research; and animal models of human disease: Pulmonary fibrosis and Hageman Trait.
Frequency: Quarterly
Availability: Registry of Comparative Pathology

Title: CRREL Benchnotes: U.S. Army Corps of Engineers Information Exchange Bulletin, May 1976- . D103.33/13:(nos.)
Producer: Cold Regions Research and Engineering Laboratory, U.S. Army Corps of Engineers, Hanover, NH 03755
Scope: Reports on the projects of CRREL researchers. Examples of projects include space shuttle icing, icing on power lines, and energy efficiency in Alaskan buildings. Also includes short news items, books of interest, and a list of recent CRREL publications.
Frequency: Bimonthly
Availability: CRREL, Hanover, NH 03755

Title: DACS Newsletter, v. 1, no. 1, October 1978- .
Producer: Data & Analysis Center for Software, RADC/ISISI, Griffis AFB, NY 13441
Scope: Contains software-related news items as well as summaries of publications of interest in the area of software acquisition, software engineering, and software languages
Frequency: Quarterly
Availability: DACS

Title: Endangered Species Technical Bulletin, v. 1, 1976- . I49.77: (v.nos.&nos.)
Producer: Endangered Species Program, U.S. Fish and Wildlife Service, Department of Interior, Washington, DC 20240
Scope: Contains articles on species of flora and fauna which are endangered or threatened. Regular features include Box Score of Species Listings (as endangered or threatened) and New Publications. Recent issues published articles on bald eagles, rare orchids, and the ocelot.
Frequency: Monthly
Availability: Endangered Species Program, depository item

Title: Enhanced Oil Recovery and Improved Drilling Technology, no. 1, 1975- . E1.28:BETC
Producer: Bartlesville Energy Technology Center, Attn: Technology Transfer Section, P.O. Box 1398, Bartlesville, OK 74005
Scope: Consists of progress reports on enhanced oil and gas recovery projects contracted out by the Department of Energy
Frequency: Quarterly
Availability: Bartlesville Energy Technology Center

Title: Environmental Response Newsletter, Summer 1976- . TD5.28/3:(date)
Producer: Commandant (G-WER-2), U.S. Coast Guard, Washington, DC 20593
Scope: This newsletter acts as a forum for exchange of ideas and information for the Coast Guard personnel involved with marine oil spills. Describes proven methods and techniques of fighting oil spills.
Frequency: Tri-annually
Availability: Coast Guard

Title: EPA Technology Transfer: The Bridge between Research and Use, July 1, 1971- . EP7.9:(date)
Producer: Center for Environmental Research Information, U.S. EPA, Cincinnati, OH 45268
Scope: Describes EPA handbooks, manuals and reports and announces conferences with the objective of disseminating EPA-developed technologies
Availability: CERI, EPA, Cincinnati, depository item

Title: FDA Drug Bulletin: Information of Importance to Physicians and Other Health Professionals, no. 1, June 1971- . HE20.4003/3: (v.nos.&nos.)
Producer: Food and Drug Administration, 5600 Fishers Lane, Rockville, MD 20857
Scope: Contains brief items on new treatments approved, potential adverse reactions of drugs, and lists of potentially hazardous products recalled by FDA
Frequency: Irregular
Availability: FDA, depository item

Title: FESA Briefs, v. 1, no. 1, February 1980- .
Producer: U.S. Army Facilities Engineering Support Agency, Fort Belvoir, VA 22060
Scope: This brief newsletter contains succinct news items of interest to engineering personnel. A recent issue, for example, describes the operation and maintenance of de-aerators and the contamination of drinking water due to backflow conditions where contaminated fluids flow into drinking water systems. It also listed new FESA publications.
Frequency: Quarterly
Availability: Army Corps of Engineers

Title: Fish Health News: A Service to the Field of Fish Health Research, v. 1, 1972- . I49.79:(v.nos.&nos.)
Producer: National Fish Health Research Laboratory, Fish and Wildlife Service, U.S. Department of the Interior, Box 700, Kearneysville, WV 25430
Scope: Publishes short reports of original research to provide up-to-date information to readers and to enable authors to document priority of their work. Includes book reviews and notices of interest to fish health researchers. Bulk of the newsletter, however, is devoted to Library Accessions which contains abstracts pertaining to fish health research.
Frequency: Quarterly
Availability: Superintendent of Documents, depository item

Title: Forestry Research West, January 1979- . A13.95:(date/nos.)
Producer: Rocky Mountain Forest and Range Experiment Station, 240 West Prospect Street, Fort Collins, CO 80526
Scope: Summarizes recent developments in forestry research conducted at the Western Experiment Stations of the USDA Forest Service. Also has a regular New Publications feature.
Frequency: Quarterly
Availability: Rocky Mountain Forest & Range Experiment Station, depository item

Title: The GACIAC Bulletin, v. 1, no. 1, September 1978- .
Producer: Guidance & Control Information Analysis Center, IIT Research Institute, Division E, 10 West Thirty-fifth Street, Chicago, IL 60616
Scope: Covers tactical weapons and their guidance and control systems such as anti-tank weapons and missiles. Announces publications and workshops and summarizes articles of interest published in other journals.
Frequency: Bimonthly
Availability: Guidance & Control Information Analysis Center

Title: Landsat Data Users Notes, no. 1, 1978- . I19.78:(nos.)
Producer: EROS Data Center, National Oceanic and Atmospheric Administration, Sioux Falls, SD 57198

Scope: Includes items of interest to the remote sensing community and specifically on Landsat. Related newsletter is *Satellite Data Users Bulletin.*
Frequency: Quarterly
Availability: NOAA Landsat Customer Services, EROS Data Center, Sioux Falls, SD 57198, depository item

Title: Liquid Fossil Fuel Technology: Quarterly Progress Report, 1978- . E1.97:(date)
Producer: Bartlesville Energy Technology Center, U.S. Department of Energy, P.O. Box 1398, Bartlesville, OK 74005
Scope: Describes research and development projects undertaken by the four divisions of the Bartlesville Energy Technology Center: Extraction, Processing, Utilization, and Project Integration and Technology Transfer Divisions.
Frequency: Quarterly
Availability: BETC

Title: Materials and Components in Fossil Energy Applications, no. 1, June 20, 1975- . E1.23:(nos.)
Producer: Division of Systems Engineering/Energy Technology, Department of Energy, Mail Stop GTN, C-157, Washington, DC 20545
Scope: It is a medium for exchange of information regarding the performance and reliability of materials and components used in the development of fossil energy systems such as coal gasification and liquefaction, and oil shale retorting. Examples of topics covered are performance of valve materials under erosive conditions and thermal distortion in air distributor plates. Also includes meeting notices and recent publications.
Frequency: Bimonthly
Availability: Wate T. Bakker, DOE, Division of Systems Engineering/Energy Technology

Title: National Cartographic Information Center (NCIC) Newsletter, v. 1, 1975- . I19.71:(nos.)
Producer: National Cartographic Information Center, U.S. Geological Survey, 507 National Center, Reston, VA 22092
Scope: Publishes news items of interest to map librarians and others interested in maps
Frequency: Irregular
Availability: National Cartographic Information Center, depository item

Title: NBS Update, October 8, 1979- . C13.36/7:(date)
Producer: Public Information Division, National Bureau of Standards, Washington, DC 20234
Scope: Provides brief information on research activities taking place in various NBS laboratories
Frequency: Biweekly
Availability: National Bureau of Standards

Title: NIAAA Information and Feature Service, v. 1, 1974- . HE20.8312:(nos.)
Producer: National Clearinghouse for Alcohol Information, P.O. Box 2345, Rockville, MD 20852
Scope: " ... highlights the activities of the NIAAA (National Institute on Alcohol Abuse and Alcoholism) and national organizations, as well as reporting on trends in programming and research across the nation."
Frequency: 12 to 14 times per year
Availability: Clearinghouse, depository item

Title: NSRDS Reference Data Report: An Informal Communication of the National Standard Reference Data System, v. 1, January/February 1977- . C13.48/3:(v.nos.&nos.)
Producer: Office of Standard Reference Data, Physics Bldg., Room A320, National Bureau of Standards, Washington, DC 20234
Scope: It is a medium of communication "for the exchange of news and ideas about data centers, publications, meetings and other activities related to data evaluation and dissemination."
Frequency: Bimonthly
Availability: Office of Standard Reference Data, NBS

Title: NTIAC Newsletter, v. 1, no. 1, april 1974- .
Producer: Nondestructive Testing Information Analysis Center, Southwest Research Institute, 6220 Culebra, P.O. Drawer 28510, San Antonio, TX 78284
Scope: This newsletter is free to public and private sector people who are engaged in nondestructive testing activities. Contents include technical articles, calls for papers for forthcoming meetings, announcements, meetings calendar, contract awards, negotiations, bid requests and price list of NTIAC publications.
Frequency: Quarterly
Availability: NTIAC

Title: NTP Technical Bulletin, v. 1, no. 1, 1979- . HE20.23/3:(nos.)
Producer: National Toxicology Program, Public Health Service, Department of Health and Human Services, Landow Building 3A-06, 7910 Woodmont Avenue, Bethesda, MD 20205
Scope: Contains news reports on various studies and activities related to the toxic effects of chemicals on animals and humans. A recent issue, for example, reported on the carcinogenic effects of

orthophthalic acid esters; summarized toxicology and carcinogenesis bioassay technical reports which were peer-reviewed; and listed the agenda for a forthcoming meeting of NTP Board of Scientific Counselors.
Frequency: Quarterly
Availability: Public Information Office, National Toxicology Program, P.O. Box 12233, MD B2-04, Research Triangle Park, NC 27709, depository item

Title: Ocean Engineering Technical Bulletin, v. 1, 1975- . C55:30:(v.nos.&nos.)
Producer: NOAA Data Buoy Office, National Oceanic and Atmospheric Administration, NSTL Station, MS 39529
Scope: Brief technical reports on ocean engineering and meteorological data buoys. Examples of topics covered are: subsurface ocean temperature measurement, measuring the directional properties of ocean waves, and ocean-atmospheric interaction study.
Frequency: Irregular
Availability: NOAA Data Buoy Office

Title: Quads: Report on Energy Activities, no. 1, 1978- . A13.98/2 (nos.)
Producer: Forest Products Laboratory, Forest Service, U.S. Department of Agriculture, P.O. Box 5130, Madison, WI 53705
Scope: Promotes the use of forest products as alternate sources of energy. Lists installations and pilot projects which use alternate sources such as fuel wood, wood gasification and pyrolysis, ethanol and other liquid fuels and densified fuel. Also reports publications, motion pictures and meetings dealing with the above alternate energy sources.
Frequency: Monthly
Availability: Forest Products Laboratory

Title: The Quarterly CERCular Information Bulletin, v. 1, no. 1, July 1976- . D103.42/11:(v.nos.&nos.)
Producer: Coastal Engineering Information Analysis Center, U.S. Army Coastal Engineering Research Center, Kingman Bldg., Fort Belvoir, VA 22060
Scope: Contains brief technical articles and news items of interest to coastal engineering research scientists. Also announces Coastal Engineering Research Center's recent publications.
Frequency: Quarterly
Availability: CERC

Title: RAC Newsletter, v. 1, no. 1, January 1970- .
Producer: Reliability Analysis Center, Rome Air Development Center, Griffiss Air Force Base, NY 13441
Scope: Announces publications and forthcoming workshops, conferences and training courses in the area of reliability analysis

Frequency: Quarterly
Availability: Reliability Analysis Center

Title: Recombinant DNA Technical Bulletin, v. 1, 1978- . HE20.3460:(v.nos.&nos.)
Producer: Office of Recombinant DNA Activities, National Institute of Allergy and Infectious Diseases, National Institutes of Health, Department of Health and Human Services, Bethesda, MD 20205
Scope: Publishes reports of progress in the recombinant DNA research. Also includes NIH activities in this area along with news and comment, announcements, and bibliography.
Frequency: Four times per year
Availability: Superintendent of Documents, depository item

Title: Research Resources Reporter, v. 1, no. 1, January 1977- . HE20.3013/6:(v.nos.&nos.)
Producer: Research Resources Information Center, 1776 East Jefferson Street, Rockville, MD 20852
Scope: Describes research in progress at various sections of the Division of Research Resources of the National Institutes of Health. Reports on the activities undertaken by Animal Resources, Biotechnology Resources, General Clinical Research Centers, Minority Biomedical Research Support, and Biomedical Research Support. Recent issues covered topics such as vaccines, intrauterine growth retardation, progress in diabetes research, bone loss in the elderly, and chimpanzee antibodies.
Frequency: Monthly
Availability: Research Resources Information Center, depository item

Title: RSIC Newsletter, v. 1, no. 1, 1963- .
Producer: Radiation Shielding Information Center, Oak Ridge National Laboratory, P.O. Box X, Oak Ridge, TN 37830
Scope: Current awareness publication. Announces recent literature, computer codes, and forthcoming calendar of events.
Frequency: Monthly
Availability: RSIC

Title: Sandia Science News, v. 1, no. 1, 1965- .
Producer: Public Information Division 3161, Sandia National Laboratories, Box 5800, Albuquerque, NM 87175
Scope: It is a 4-page newsletter reporting on the research and development projects in progress at the laboratories. Examples of these projects are lead implantation technique which simulates nuclear waste damage and the development of seabed penetrator prototypes that can be used to gather marine sediment data. Articles are brief and are understandable to non-technical people.
Availability: Sandia National Laboratories

Title: Science Resources Studies Highlights, 1969- . NS1.31:(nos.)
Producer: Division of Science Resources Studies, National Science Foundation, Washington, DC 20550
Scope: Production and employment picture of science and engineering graduates
Availability: National Science Foundation

Title: Sea Grant Today, v. 1, no. 1, September 1970- .
Producer: Extension Division, Virginia Polytechnic Institute and State University, Blacksburg, VA 24061
Scope: This publication is supported by the National Oceanic and Atmospheric Administration. Articles are in popular language and deal with economic, legal and recreational aspects of ocean. Each issue also lists an extensive list of publications resulting from National Sea Grant College Program. These publications are listed under subjects such as aquaculture, biology, commercial fishing, diving, ecology, estuarine studies, ocean engineering, and pollution.
Frequency: Bimonthly
Availability: Virginia Polytechnic

Title: SEAN (Scientific Event Alert Network) Bulletin, January 31, 1978- .
Producer: SEAN, National Museum of Natural History, MRC 129, Smithsonian Institution, Washington, DC 20560
Scope: Reports on transient geophysical and biological events such as volcanic events, earthquakes, and meteoritic events
Frequency: Monthly
Availability: American Geophysical Union, 2000 Florida Avenue, NW, Washington, DC 20009

Title: Smithsonian Institution Research Reports, v. 1, 1972- . SI1.37: (nos.)
Producer: Office of Public Affairs, Smithsonian Institution, Washington, DC 20560
Scope: Contains news stories on Smithsonian Institution's research activities. Covers astronomy, marine ecology, biogeology, botany and astrophysics among others.
Frequency: Three times per year
Availability: Smithsonian Institution

Title: Technology News, no. 1, February 1974- . I28.152:(nos.)
Producer: Technology Transfer Group, Bureau of Mines, 2401 E Street, NW, Washington, DC 20241
Scope: Describes "tested developments" from the Bureau of Mines Research Programs and is published to "encourage the transfer of this information to the minerals industry and its application in commercial practice." Each 2-page issue describes a machine or a process under headings such as objective, approach, how it works, test results, patent status, and for more information.

Frequency: Irregular
Availability: Technology Transfer Group

Title: Tech-Tran, v. 1, 1976- .
Producer: U.S. Army Engineer Topographic Laboratories, Fort Belvoir, VA 22060
Scope: This newsletter describes the activities and achievements of the Topographic Laboratories. The laboratory is active in digitizing geographic and cartographic data for both civilian and military use. Each issue lists the reports and papers published by the Topographic Laboratories.
Frequency: Quarterly
Availability: U.S. Army Engineer Topographic Laboratories

Title: Thermophysics and Electronics Newsletter, v. 1, no. 1, January/February, 1972- .
Producer: Thermophysical and Electronic Properties Information analysis Center, TEPIAC/CINDAS, Purdue University, West Lafayette, IN 47907
Scope: Areas of interest include numerical data in the area of thermophysical and electronic properties of materials
Frequency: Quarterly
Availability: TEPIAC

Title: TSCA Chemical-in-Progress Bulletin, v. 1, 1980- . EP5.15: (v.nos.&nos.)
Producer: Industry Assistance Office (TS-799), Office of Pesticides & Toxic Substances, U.S. Environmental Protection Agency, Washington, DC 20460
Scope: This newsletter reports the regulatory activities of the Environmental Protection Agency in relation to the Toxic Substances Control Act (TSCA). Includes regulatory actions, lists of chemicals that are going to be manufactured or imported, and EPA actions regarding hazardous substances.
Frequency: Bimonthly
Availability: Industry Assistance Office, EPA, depository item

Title: Upper Atmospheric Programs Bulletin, January 1976- . NAS1.63:(nos.)
Producer: Office of Environment and Energy (AEE-300), Federal Aviation Administration, 800 Independence Avenue, SW, Washington, DC 20591
Scope: The *Bulletin* is funded and published jointly by the Upper Atmospheric Research Program of NASA and High Altitude Pollution Program of FAA. Contains news lighlights, summaries of publications of interest, abstracts of papers, announcements of forthcoming events and lists of recent reports.
Frequency: Bimonthly
Availability: FAA

Title: Water Impacts, v. 1, January 1980- .
Producer: Institute of Water Research, 334 Natural Resources Bldg., Michigan State University, East Lansing, MI 48824
Scope: The Institute of Water Research is partially funded by the U.S. Department of the Interior. *Water Impacts* consists of news items on water quality, resources and technology. Also carries lists of conferences, meetings, short courses, and publications.
Frequency: Monthly
Availability: Institute of Water Research

Title: Water Resources Center, University of Delaware News, no. 1, October 1971- .
Producer: Water Resources Center, University of Delaware, Newark, DE 19711
Scope: Contains brief articles on hydrology and water resources
Frequency: Irregular
Availability: Delaware Water Resources Center

EMPLOYEE-ORIENTED PERIODICALS

Title: ADAMHA News, v. 1, 1975- . HE20.8013:(v.nos.&nos.)
Producer: ADAMHA Office of Communications and Public Affairs, Alcohol, Drug Abuse, and Mental Health Administration, Room 12C-15, 5600 Fishers Lane, Rockville, MD 20857
Scope: Reports ADAMHA-related news such as federal strategy on drug abuse, effects of marijuana on adolescents, link between mood disorder and menstrual cycle, and programs to aid the deinstitutionalized patients
Frequency: Biweekly
Availability: ADAMHA, depository item

Title: Air Defense, 1976- . D101.77:(date)
Producer: U.S. Army Air Defense School, ATTN: ATSA-TLD-S, Fort Bliss, TX 79916
Scope: Informs the air defense personnel of the developments in tactical, doctrinal, and technical developments in air defense field. Topics such as covert/passive air defense sensors, Soviet helicopters and electromagnetic pulse associated with nuclear weapon detonation are covered.
Frequency: Quarterly
Availbility: Custodian, U.S. Army Air Defense Magazine Fund, USAADS, ATTN: ATSA-TDL-S, Fort Bliss, TX 79916, depository item

Title: Air Force Engineering and Services Quarterly, v. 1, no. 1, February 1960- . D301.65:(v.nos.&nos.)
Producer: HQ Air Force Engineering and Services Center, Tyndall AFB, FL 32403
Scope: Publishes articles that provide "current information about significant civil engineering and services programs and accomplishments throughout the Air Force" as well as "pertinent developments

in science and engineering.... " Published contributions deal with topics such as MHD generators, R&D in environmental quality and civil engineering, and construction cost overruns. Also has a number of regular features such as Tecnotes and People Worth Noting.
Frequency: Quarterly
Availability: Superintendent of Documents, depository item

Title: Airman, v. 1, 1957- . D301.60:(v.nos.&nos.)
Producer: Air Force Service Information and News Center (AFSINC), Kelly AFB, TX 78241
Scope: It is a medium of information to air force personnel. Apart from the usual morale boosting features, it also publishes informative articles on topics such as bombers and planetariums.
Frequency: Monthly
Availability: Superintendent of Documents, depository item

Title: Approach: The Naval Aviation Safety Review, v. 1, 1955- . D202.13:(v.nos.&nos.)
Producer: Naval Safety Center, Safety Publications Dept., NAS Norfolk, VA 53511
Scope: Promotes safe aviation practices. Deals with hazardous situations such as wind turbulence, helicopter inflight icing, and safety rule violations.
Frequency: Monthly
Availability: Superintendent of Documents, depository item

Title: Armor: The Magazine of Mobile Warfare, v. 1, March 1888- . D101.78/2:(v.nos.&nos.)
Producer: U.S. Army Armor School, 4401 Vine Grove Road, Fort Knox, KY 40121
Scope: Deals with ground warfare technology and tactics. *Armor* covers topics such as tanks, gunnery training, specifications of guns, and armor technology.
Frequency: Bimonthly
Availability: U.S. Armor Association, P.O. Box O, Fort Knox, KY 40121

Title: Army Communicator: Voice of the Signal Corps, v. 1, no. 1, Winter 1976- . D111.14:(v.nos.&nos.)
Producer: U.S. Army Signal Center, Fort Gordon, GA 30905
Scope: Publishes doctrinal and technical information on communications and electronics. Presents articles such as satellite communications, the role of Signal Corps in the U.S. space program, and the role of communications in battle field.
Frequency: Quarterly
Availability: TAC Subscription Fund, U.S. Army Signal Center, Fort Gordon, GA 30905

Title: Army Logistician, v. 1, no. 1, September/October 1969- . D101.69:(v.nos.&nos.)
Producer: U.S. Army Logistics Management Center, Fort Lee, VA 23801
Scope: Aims to inform the military and civilian employees of the Army on defense logistics
Frequency: Bimonthly
Availability: Superintendent of Documents, depository item

Title: Army RD&A: Army Research, Development & Acquisition Magazine, v. 1, no. 1, December 1960- . D101.52/3:(v.nos.&nos.)
Producer: Development, Engineering and Acquisition Directorate (DRCDE), HQ U.S. Army Materiel Development and Readiness Command (DARCOM), Alexandria, VA 22333
Scope: Source of information on Army's RD&A activities and personnel
Frequency: Bimonthly
Availability: Superintendent of Documents, depository item

Title: ARRADCOM Voice, v. 1, no. 1, June 1977- .
Producer: Media Unlimited, P.O. Box 53, Mt. Arlington, NJ 07856
Scope: It is a newspaper published by a private firm for the employees of the U.S. Army Armament Research and Development Command. Along with sports and advertisements, it reports on the researchers and research activities of USAARDC.
Frequency: Biweekly
Availability: Media Unlimited

Title: Clinch River Breeder Reactor Plant Project: Breeder Briefs, April 1974- .
Producer: Breeder Reactor Corporation, P.O. Box U, Oak Ridge, TN 37830
Scope: Reports on developments in the Clinch River Project
Frequency: Monthly
Availability: Breeder Reactor Corporation

Title: Crime Laboratory Digest, v. 1, no. 1, March 1974- .
Producer: FBI Laboratory Division, Federal Bureau of Investigation, Washington, DC 20535
Scope: It is an informal means of communication among crime laboratory scientists.
Frequency: Every two to three months
Availability: FBI

Title: Defense, v. 1, 1978- . D2.15/3:(date)
Producer: American Forces Information Service, 1735 North Lynn Street, Room 210, Arlington, VA 22209
Scope: Addressed to commanders and key personnel, this magazine emphasizes defense policies and programs. Recent articles included chemical warfare, nuclear freeze, quality control in defense industries, and the role of military in basic scientific research.

Frequency: Monthly
Availability: Superintendent of Documents, depository item

Title: Driver: The Traffic Safety Magazine for the Military Driver, v. 1, 1957- . D301.72:(v.nos.&nos.)
Producer: HQ Air Force Inspection and Safety Center, Department of the Air Force, Norton Air Force Base, CA 92409
Scope: It is an informal magazine designed to provide safety tips to motor vehicle drivers.
Frequency: Monthly
Availability: Superintendent of Documents, depository item

Title: Energy Insider, v. 1, 1978- . E1.54:(v.nos.&nos.)
Producer: Office of Public Affairs, Department of Energy, Washington, DC 20585
Scope: Contains news items regarding the Department of Energy
Frequency: Monthly
Availability: Superintendent of Documents, depository item

Title: Engineer: The Magazine for Army Engineers, v. 1, no. 1, Spring 1971- . D103.115:(v.nos.&nos.)
Producer: U.S. Army Engineer Center and Fort Belvoir, VA, Fort Belvoir, VA 22060
Scope: Focuses on combat engineering. Covers topics such as critical path method, fabrics in military construction, and engineer workbooks.
Frequency: Quarterly
Availability: Superintendent of Documents, depository item

Title: Engineer Update, v. 1, 1977- . D103.69:(v.nos.&nos.)
Producer: U.S. Army Corps of Engineers (HQ USACE), DAEN-PAR, Washington, DC 20314
Scope: Primarily directed to the employees of the corps. Publishes short news items accompanied by photographs.
Frequency: Monthly
Availability: Editor, *Engineer Update*, depository item

Title: FAA General Aviation News: A DOT/FAA Flight Operations Safety Publication, v. 1, no. 1, May 1962- . TD4.9:(v.nos.&nos.)
Producer: Flight Operations, AFO-807, Federal Aviation Administration, Department of Transportation, Washington, DC 20591
Scope: It is designed to inform the general aviation airmen of the technical, regulatory and procedural matters affecting the operation of aircraft. Covers topics such as freshwater flying, fuel tank venting, and survival at sea.
Frequency: Bimonthly
Availability: Superintendent of Documents, depository item

Title: Faceplate: The Official Magazine for the Divers and Salvors of the United States Navy, v. 1, no. 1, 1971- . D211.22:(v.nos.&nos.)
Producer: Supervisor of Diving (SEA OOC-D), Naval Sea Systems Command, Department of the Navy, Washington, DC 20362
Scope: Brings "the latest and most informative news available to the Navy diving and salvage community.... "
Frequency: Quarterly
Availability: Superintendent of Documents, depository item

Title: Fathom: Surface Ship & Submarine Safety Review, 1969- . D202.20:(v.nos.&nos.)
Producer: Naval Safety Center, NAS Norfolk, VA 23511
Scope: Aims to improve the safety on ships and submarines. Examples of topics covered include: mishaps caused by synthetic lines, health hazards associated with fire-resistant hydraulic fluids, electric shocks from workbenches, static electricity, and dangers of halocarbons.
Frequency: Quarterly
Availability: Superintendent of Documents, depository item

Title: Fish and Wildlife News, November 1973- . I49.88:(date)
Producer: Office of Current Information, Public Affairs, U.S. Fish and Wildlife Service, Department of the Interior, Washington, DC 20240
Scope: It is published for the employees of the Fish and Wildlife Service. It includes news items from the seven regions of the service. These news items and other features generally deal with conservation of wildlife.
Frequency: Bimonthly
Availability: U.S. Fish and Wildlife Service, depository item

Title: Flying Safety, v. 1, 1945- . D301.44:(v.nos.&nos.)
Producer: Department of the Air Force, HQ Air Force Inspection and Safety Center, Norton Air Force Base, CA 92409
Scope: Emphasizes safe flying and prevention of mishaps. Articles tend to describe accidents and near-accidents and analyzing their causes.
Frequency: Monthly
Availability: Superintendent of Documents, depository item

Title: LBL News Magazine, v. 1, 1976- . E1.53/2:(v.nos.&nos.)
Producer: Lawrence Berkeley Laboratory, University of California, Berkeley, CA 94720
Scope: Reports on work in progress in accelerator and fusion research, nuclear science, physics, computer science and mathematics, chemical biodynamics, earth sciences, energy and environment, and materials and molecular research
Frequency: Quarterly
Availability: Lawrence Berkeley Laboratory, depository item

Title: Lifesaver: The U.S. Coast Guard Safety and Occupational Health Review, v. 1, no. 1, Winter 1980- . TD5.42:(v.nos.&nos.)
Producer: Safety Programs Division, U.S. Coast Guard Headquarters, Washington, DC 20593
Scope: Provides safety information to Coast Guard personnel. Topics covered include engine room fire safety, non-ionizing radiation, and safety and health inspections.
Frequency: Quarterly
Availability: Safety Programs Division, U.S. Coast Guard

Title: Mech: The Naval Aviation Maintenance Safety Review, v. 1, 1968- . D202.19:(date)
Producer: Naval Safety Center, Norfolk, VA 23511
Scope: Presents "a broad spectrum of aviation maintenance matters dealing with such topics as maintenance-caused mishaps, personnel/material hazards, maintenance management/safety programs, safety awareness, etc., as well as material on aviation ground safety in general." Examples of articles include FOD (Foreign Object Damage), troubleshooting of F-14 missile release system, and safe handling of beryllium brake assemblies.
Frequency: Quarterly
Availability: Superintendent of Documents, depository item

Title: Military Intelligence, 1974- . D101.84:(v.nos.&nos.)
Producer: U.S. Army Intelligence Center and School, Fort Huachuca, AZ 85613
Scope: Intelligence doctrine, policy, and counter-intelligence are covered in this magazine. Examples of topics published include laser weapons, enigma cipher machine, chemical warefare, tactical intelligence, and reconnaissance.
Frequency: Quarterly
Availability: Superintendent of Documents, depository item

Title: Mine Safety & Health, v. 1, 1975- . L38.9:(v.nos.&nos.)
Producer: Office of Information, Mine Safety and Health Administration, 4015 Wilson Blvd., Room 604, Ballston Tower No. 3, Arlington, VA 22203
Scope: Aims to inform and educate the mining community on communications and safety, mine rescue, and safety training. Also includes a section called Safety Technology which lists equipment approved by the Mine Safety and Health Administration.
Frequency: Quarterly
Availability: Superintendent of Documents, depository item

Title: Naval Aviation News, October 1, 1919- . D202.9:(date)
Producer: Chief of Naval Operations and Naval Air Systems Command, Bldg. 146, Washington Navy Yard, Washington, DC 20374

Scope: Provides professional information to naval aviation community. Articles deal with topics such as use of computers in flight simulation and search and rescue, drug abuse, and the role of navy in space flights.
Frequency: Monthly
Availability: Superintendent of Documents, depository item

Title: *The Navigator*, v. 1, no. 1, Summer 1953- . D301.38/4:(v.nos.&nos.)
Producer: Headquarters 323D Flying Training Wing (ATC), Mather Air Force Base, CA 95655
Scope: Publishes brief articles on celestial navigation, navigation techniques, weapons systems, training, and professional development
Frequency: Three times per year
Availability: Superintendent of Documents, depository item

Title: *Navy Civil Engineer*, v. 1, no. 1, February 1960- . D209.13:(v.nos.&nos.)
Producer: Navy Civil Engineer, Code C20M, Naval School, Civil Engineer Corps Officers, Port Hueneme, CA 93043 (Published by: Naval Facilities Engineering Command, Alexandria, VA 22332)
Scope: Designed to inform the personnel of U.S. Navy Civil Engineer Corps. Contains articles on topics such as construction, planning and design, and public works.
Frequency: Quarterly
Availability: Superintendent of Documents, depository item

Title: *Navy Lifeline: Safety and Occupational Health Journal*, v. 1, 1972- . D207.15:(v.nos.&nos.)
Producer: Naval Safety Center, Code 70, Norfolk, VA 23511
Scope: Informs and educate naval personnel on all aspects of occupational health hazards resulting from hazardous materials, electrical equipment, fires, motor vehicles and others such as lawn mowers and chain saws
Frequency: Bimonthly
Availability: Superintendent of Documents, depository item

Title: *The NIH Record*, 1949- .
Producer: National Institutes of Health, Public Health Service, Bldg. 31, Room 2B-03, Bethesda, MD 20205
Scope: Publishes brief items of interest to NIH employees. Includes news, interviews, availability of training courses and the like.
Frequency: Biweekly
Availability: Editorial Operations Branch, Division of Public Information, NIH

Title: *Proceedings of the Marine Safety Council*, v. 1, 1944- . TD5.13:(v.nos.&nos.)
Producer: U.S. Coast Guard, 2100 Second Street, SW, Washington, DC 20593

Scope: Designed to improve safety at sea, the *Proceedings* publishes readable articles on oil spills, marine regulations, hazardous chemicals, news, and personnel matters.
Frequency: Monthly
Availability: Coast Guard, depository item

Title: Program Manager: The Defense Systems Management College Newsletter, v. 1, no. 1, January/February 1972- . D1.60:(v.nos.& nos.)
Producer: Defense Systems Management College, Fort Belvoir, VA 22060
Scope: Contains brief articles and news items on defense systems acquisition
Frequency: Bimonthly
Availability: Defense Systems Management College, depository item

Title: United States Army Aviation Digest, February 1955- . D101.47:(v.nos.&nos.)
Producer: U.S. Army Aviation Center, ATTN: ATZQ-ES-AD, Fort Rucker, AL 36362
Scope: Provides information on aircraft operation and safety, training, maintenance, operations, research and development, and aviation medicine
Frequency: Monthly
Availability: Superintendent of Documents, depository item

Title: World, v. 1, 1971- . TD4.9/2:(v.nos.&nos.)
Producer: Public & Employee Communications Division, Office of Public Affairs, Federal Aviation Administration, 800 Independence Avenue, SW, Washington, DC 20591
Scope: Contains articles on technical aspects of FAA operations. Recent issues described aircraft simulators and solid-state and remote monitoring equipment.
Frequency: Monthly
Availability: FAA

PUBLIC INFORMATION JOURNALS

Title: Agricultural Research, v. 1, no. 1, January/February 1953- . A106.9:(v.nos.&nos.)
Producer: Information Staff, Room 3145-S, Agricultural Research Service, U.S. Department of Agriculture, Washington, DC 20250
Scope: Communicates to the public the results of research sponsored by ARS. The brief and non-technical articles are published under the headings: crop sciences, livestock and veterinary sciences, post-harvest science and technology, genetic engineering in agriculture and soil, water, and air sciences.
Frequency: Monthly
Availability: Superintendent of Documents, depository item

Title: Argonne News: About the People and Programs of Argonne National Laboratory, v. 1, no. 1, 1953- . E1.86/2:(v.nos.&nos.)
Producer: Office of Public Affairs, Argonne National Laboratory, 9700 South Cass Avenue, Argonne, IL 60439
Scope: Publishes news items on the research in progress at Argonne National Laboratory. Articles cover topics such as pulsed neutron research, use of gold in materials research, and modification of hodoscope which allows scientists to examine the core of a nuclear reactor.
Availability: Office of Public Affairs, Argonne National Laboratory, depository item

Title: Changing Scene, 1959- .
Producer: Ames Laboratory, Iowa State University, Ames, IA 50011
Scope: Each issue describes one or two research efforts in progress at Ames Laboratory. Research areas include low temperature physics and thermometry, and use of microprocessors in designing scientific instruments.
Frequency: Monthly
Availability: Ames Laboratory

Title: Earthquake Information Bulletin, v. 1, no. 1, 1967- . I19.65:(v.nos.&nos.)
Producer: Geological Survey, Department of the Interior, Reston, VA 22092
Scope: Provides current information on earthquakes and seismological activities of interest to the general public. Includes Questions and Answers, announcements, lists of new publications, and summaries of all earthquakes felt throughout the world during the preceding two-month period.
Frequency: Bimonthly
Availability: Superintendent of Documents, depository item

Title: EPA Journal, v. 1, 1975- . EP1.67:(v.nos.&nos.)
Producer: Office of Public Affairs (A-107), Environmental Protection Agency, Washington, DC 20460
Scope: Describes the policies and programs of the Environmental Protection Agency. Deals with topics such as hazardous wastes, recycling wastes and other related areas.
Frequency: Bimonthly
Availability: Superintendent of Documents, depository item

Title: E&TR: Energy and Technology Review, April 1975- . E1.53:(date)
Producer: Technical Information Department, Lawrence Livermore Laboratory, University of California, Livermore, CA 94550
Scope: The review reports the laboratory's accomplishments in its unclassified programs. The laboratory conducts research in magnetic fusion energy; laser isotope separation; chemistry, engineering, and physics; biomedical and environmental sciences; and

applied energy technology. *Research Monthly* is a companion periodical that reports on the laboratory's classified research.
Frequency: Monthly
Availability: National Technical Information Service, depository item

Title: FDA Consumer, v. 1, no. 1, February 1967- . HE20.4010:(v.nos.&nos.)
Producer: Food and Drug Administration, HFI-20, 5600 Fishers Lane, Rockville, MD 20857
Scope: Designed to inform the consumer of foods, drugs, and cosmetics. Articles are written in non-technical language.
Frequency: Monthly except for July-August and December-January combined issues
Availability: Superintendent of Documents, depository item

Title: Fermilab Report, 1970- . E1.92/2:(nos.)
Producer: Fermi National Accelerator Laboratory, P.O. Box 500, Batavia, IL 60510
Scope: Brief notes, with accompanying photographs, on nuclear accelerator research
Frequency: Monthly
Availability: Fermi Laboratory, depository item

Title: Food News for Consumers, January 1980- . A110.10:(date)
Producer: Food Safety and Inspection Service, U.S. Department of Agriculture, Washington, DC 20250
Scope: Consumer-oriented publication with a variety of news items such as safe cooking of pork, ingredients in processed foods, tips on freezing meat and poultry, and food stamps
Frequency: Four issues per year
Availability: Superintendent of Documents, depository item

Title: Information Bulletin, 1976- .
Producer: Communications Office, Princeton University, Plasma Physics Laboratory, James Forrestal Research Campus, P.O. Box 451, Princeton, NJ 08544
Scope: Brief, readable bulletins explaining different aspects of PPL and fusion research
Frequency: Irregular
Availability: Plasma Physics Laboratory

Title: Logos: Progress through Science, November 1982- .
Producer: Argonne National Laboratory, 9700 South Cass Avenue, Argoone, IL 60439
Scope: Provides "updated reports on selected programs at Argonne National Laboratory"
Frequency: Irregular
Availability: Argonne National Laboratory

Title: Los Alamos Science, v. 1, no. 1, 1980- .
Producer: Los Alamos National Laboratory, Mail Stop M708, Los Alamos, NM 87545
Scope: Reports on research taking place at the laboratory. Articles, which are accompanied by graphics, are oriented to educated laypersons. Carries features such as interviews, editor's note, and science ideas. Recent articles dealt with cosmic gamma-ray bursts and nuclear microprobes which enable the study of surfaces by means of ions.
Frequency: Quarterly
Availability: Los Alamos National Laboratory

Title: Mosaic, v. 1, no. 1, Winter 1970- . NS1.29:(v.nos.&nos.)
Producer: National Science Foundation, Washington, DC 20550
Scope: Reports on scientific research supported by the National Science Foundation
Frequency: Bimonthly
Availability: Superintendent of Documents, depository item

Title: News & Features from NIH HE20.3007/2:(date)
Producer: NIH News Branch, Bldg. 31, Room 2B-10, Bethesda, MD 20205
Scope: Covers NIH or NIH-supported research and educational programs. Examples of topics published include Lesch-Nyhan Syndrome, cell membranes, and electron microscope.
Frequency: Bimonthly
Availability: NIH News Branch, depository item

title: NOAA, v. 1, no. 1, January 1971- . C55.14:(v.nos.&nos.)
Producer: Office of Public Affairs, National Oceanic and Atmospheric Administration, Rockville, MD 20852
Scope: Informs and educates the public regarding the programs of NOAA whose responsibilities include weather and prediction, deep-sea fishing and other aspects of marine fisheries and oceanographic and meteorological research. Also includes seafood recipes.
Frequency: Quarterly
Availability: Superintendent of Documents, depository item

Title: NSF Bulletin, v. 1, no. 1, 1973- . NS1.3:(v.nos.&nos.)
Producer: National Science Foundation, Washington, DC 20550
Scope: Announces NSF programs, publications, personnel, and other news items reflecting on NSF policies
Frequency: Monthly (except July and August)
Availability: National Science Foundation, depository item

Title: Oak Ridge National Laboratory Review, v. 1, no. 1, Summer 1967- .
Producer: Bldg. 4500-North, Oak Ridge National Laboratory, P.O. Box X, Oak Ridge, TN 37830

Scope: Source of information on the ORNL research program. Reports on research in areas such as energy, basic sciences, nuclear medicine, chemicals, and materials research.
Frequency: Quarterly
Availability: Oak Ridge National Laboratory

Title: PPL Digest, 1980- .
Producer: Information Services, Princeton Plasma Physics Laboratory, P.O. Box 451, Princeton, NJ 08544
Scope: Describes the magnetic fusion energy research taking place at the Plasma Physics Laboratory
Frequency: Quarterly
Availability: Plasma Physics Laboratory

Title: Reclamation Era: A Water Review Quarterly, v. 1, 1908- . I27.5:(v.nos.&nos.)
Producer: Bureau of Reclamation, Department of the Interior, Code D-140, P.O. Box 25007, Denver Federal Center, Denver, CO 80225
Scope: Contains articles on various aspects of water use such as irrigation, recreation, power generation, and fish and wildlife enhancement
Frequency: Quarterly
Availability: Superintendent of Documents, depository item

Title: Water Spectrum, v. 1, 1949- . D103.48:(v.nos.&nos.)
Producer: Water Resources Support Center, U.S. Army Corps of Engineers, Fort Belvoir, VA 22060
Scope: Discusses water resources issues such as legislation, water quality, flood risk and control, and environment
Frequency: Quarterly
Availability: Superintendent of Documents, depository item

Title: Your Public Lands, v. 1, 1951- . I53.12:(v.nos.&nos.)
Producer: Office of Public Affairs, Bureau of Land Management, Department of the Interior, Eighteenth and C Streets, NW, Washington, DC 20240
Scope: Communicates to the public about legislation, land resources, recreational opportunities, and other topics
Frequency: Quarterly
Availability: Superintendent of Documents, depository item

6

PATENTS

The purpose of a United States patent is to give an inventor exclusive rights protecting his invention, which is then made freely accessible to the public. A result of this mutually beneficial exchange is one of the world's greatest repositories of scientific information. In fact, the constantly growing patent file of the U.S. Patent and Trademark Office has been described as a national resource. This patent file includes over four million U.S. patents, supplemented by foreign patents and other technical publications, all contributing to a unique collection that grows by over a half million documents annually.

WHAT IS A PATENT?

A patent protects the inventor's right to prohibit others from making, using, or selling his invention in the U.S. for a period of seventeen years. Patents are granted for new and useful processes, machines, manufactures, or compositions of matter, new and distinct varieties of plants or organisms or new designs for articles of manufacture. These categories of invention are reflected in the four categories in which patents may be issued: 1) utility patents, the largest category; 2) plant patents; 3) design patents; and 4) reissue patents, which show corrections for patents previously issued.

USES OF PATENT LITERATURE

There are many needs that can be satisfied through a patent search. One of the more common is a search by a hopeful inventor to determine if a potential invention has already been patented before filing his own application with the patent office. Other people search the patent literature seeking patents for which the seventeen-year protection has expired, allowing the invention to enter the public domain.

Patents are also a valuable primary information source for tracing the history of technology, or seeking a solution for a technical problem. Examination of patents allows access to a wealth of technical information, whether approached as a general overview of technology or as related to specific categories of inventions.

> Not only does the [U.S. Patent and Trademark Office] file embody the most comprehensive collection of technical information of its kind in the world, the information is inherently presented in a manner such that nearly every significant development in almost all technical fields flows in a natural time-series sequence—virtually welcoming monitoring and analysis.[1]

Patent searches can also supplement current awareness services since a patent may be the first published form in which new information is released. Patents provide a primary information source which is current, concise, and thorough. Not only can patents provide a scientist with new insights on his own research, but also clues to the directions being pursued by competitors.

Despite the unique information available in patent literature, it is often overlooked as a resource of technical information. One reason is researchers' unfamiliarity with patent literature. Especially in current awareness searching, journals are often relied upon as a complete source of up-to-date technical information because the researchers are more familiar with the form of literature. Yet journals can lag significantly behind patents in reporting many innovations.

In addition to currency, the patent literature offers other advantages. Maynard has pointed out that, unlike technical journal literature, patents offer the advantage of presuming little background knowledge on the part of the reader.[2] Each patent must stand on its own, and offers certain common categories of information in a format that can be mastered easily.

INFORMATION PROVIDED IN A PATENT

The typical utility patent has three parts: 1) a cover page, providing basic information; 2) drawings; and 3) the patent specification, or written description of the invention. These components combine to offer a discussion of the field of technology, summary of the prior art, benefits of the discovery, a technical definition of the invention, detailed description of the invention as summarized in the definition, statements concerning its usefulness, working examples, and claims to exclusive aspects of the invention.[3]

An example of a patent cover page is shown in figure 6.1. The numbers in brackets, called INID Codes, were developed by the Paris Union committee for

FIGURE 6.1
EXAMPLE OF PATENT COVER PAGE

United States Patent [19] | [11] **4,061,241**
Retelny | [45] **Dec. 6, 1977**

[54] **FOOD PLATE PACKAGE**

[75] Inventor: **Andrew G. Retelny**, Wheaton, Ill.

[73] Assignee: **McDonald's Corporation**, Oak Brook, Ill.

[21] Appl. No.: **727,293**

[22] Filed: **Sept. 27, 1976**

[51] **Int. Cl.**[2] **B65D 1/34**; B65D 43/10; A45C 11/20

[52] **U.S. Cl.** .... **220/4 B**; 206/508; 206/511; 206/545; 220/306; 229/2.5 R; 229/43

[58] **Field of Search** .... 220/4 B, 4 E, 306; 229/2.5, 43; 206/508, 511, 512, 540, 541, 545, 549

[56] **References Cited**

U.S. PATENT DOCUMENTS

Re. 28,720	2/1976	Sedlak	206/508
3,244,311	5/1966	Lawson	220/4 B
3,303,964	2/1967	Luker	206/508
3,557,995	1/1971	Mirasol	206/508
3,613,933	10/1971	Pilz	206/545

Primary Examiner—George E. Lowrance
Attorney, Agent, or Firm—Robert E. Wagner; Robert E. Browne; Gerald T. Shekleton

[57] **ABSTRACT**

A food plate package having a bottom plate and a top cover for the temporary storage of prepared food products, the top cover being convex-oval, with support ribs integrally formed in the cover. The support ribs provide a raised flat surface to stably support another food plate package when stacking one on top of the other. Channels are provided between the support ribs for an air space between adjacent packages when stacked and the subsequent dissipation of overly high temperatures in the central portion.

9 Claims, 9 Drawing Figures

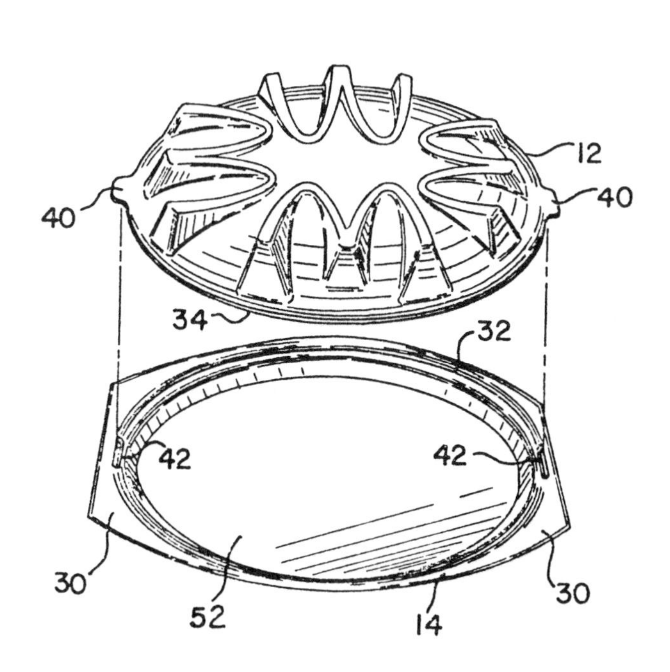

International Cooperation in Information Retrieval Among Patent Offices (ICIREPAT). Their purpose is to provide a means for identifying data on the patent cover page, even when the reader is unfamiliar with the language used or the patent laws applied.

As illustrated in the example in figure 6.1, the cover page of the patent provides vital information such as the unique patent number, name of inventor and assignee, classification, an abstract, and a drawing, when applicable. Figure 6.2 identifies some of these data elements on the example by their INID codes.

FIGURE 6.2
PATENT INID CODES

INID Code	*Data Element*
[11]	Patent Number
[45]	Patent Publication Date
[54]	Title of Invention
[21]	Application Number
[51]	International Patent Classification
[52]	U.S. Patent Classification

Following the cover page are the drawings and the patent specification. The general arrangement of the specification is usually as follows:

1. *Reference to related applications*—the relationship between this patent and earlier ones by the same inventor on the same subject.
2. *Background*—describes the field of invention and prior state of the art.
3. *Summary*—nature, operation, and purpose of the invention.
4. *Description of the drawings.*
5. *Detailed description of the invention*—sufficient explanation so that anyone skilled in the particular art could make and use the invention, and gives the best mode of carrying out the invention.
6. *Claims*—one-sentence legal definitions that specify the subject matter to which the inventor feels his invention applies. Each patent will have at least one claim, and these claims determine the legal protection afforded by the patent.[4]

PATENT CLASSIFICATION

Since the first U.S. patent was granted in 1790, over four million patents have been issued. To facilitate access to this constantly expanding body of knowledge, patents have been classified into about three hundred and fifty broad technical categories (classes), and approximately one hundred and twelve thousand specific technical groupings (subclasses). As in any classification system, patent classes and subclasses function to group together similar processes and inventions.

The primary consideration in classifying patents as members of specific classes and subclasses is utility. The concept of utility emphasizes the direct function, effect, or product of an invention, rather than its obscure or accidental uses or applications. The aim is to place inventions seeking similar results through application of similar scientific principles together in classes and subclasses. Cross referencing is performed when a patent contains significant technology pertinent to more than one subclass. In these cases, copies of the patent are placed in each of the appropriate subclass files.

The U.S. Patent Classification System forms the framework for searching for inventions of general types, when no patent number or inventor's name is known. Each class and subclass is identified by unique numbers. The *Manual of Classification*, the *Index to Classification*, and *Classification Definitions* (each of which will be discussed later) use these class and subclass numbers as access points for searching.

PATENT SEARCHING

PRINT SOURCES–U.S. PATENT OFFICE

When searching manually for patents in a technological field or field of science, knowledge and coordination of three publications of the patent office is essential: the *Index to Classification*, the *Manual of Classification*, and *Classification Definitions*. The *Index to Classification* (C21.12/2:(date)) serves as an initial point of entry when trying to define a search in terms of specific classes and subclasses.

The index is an alphabetical subject heading list, which refers the user to the classes and subclasses related to each index term. If, for example, one seeks patents for baseball gloves, one would search the broad term GLOVE in the index and find subheading Baseball under it (figure 6.3, page 94). The index shows that Baseball gloves are assigned the Class and Subclass numbers 2 and 19, respectively.

The *Manual of Classification* (C21.12:(date)) is a list of all class and subclass numbers and their descriptive titles (figure 6.4, page 95). Since it is meant to serve as a list only, the descriptive titles are brief and not definitive. To locate baseball gloves in this source, the searcher must first know (or guess at) the class. In Class 2, APPAREL, we find the listing for baseball gloves: "APPAREL, GUARDS AND PROTECTORS, Hand or arm, Baseball gloves." This breakdown demonstrates the need to construct a search based on the item's function or purpose, rather than everyday language used to describe it. The indentions in the manual indicate levels of subclassification, with each level providing greater detail.

Since the manual provides only brief descriptive statements on each class and subclass, the searcher often will use it as a key to *Classification Definitions*. After using the manual to identify the class and subclass sought, the searcher can turn to the same class and subclass listing in *Classification Definitions* for elaboration. *Classification Definitions* provides class and subclass listings, followed by notes and definitions meant to clarify inclusions and exclusions for each category.

FIGURE 6.3
PATENT INDEX CLASSIFICATION

GLASS (Continued)

	Class	Subclass
Bead		
Reflector	88	78+
Reflector making	117	16+
Block construction	52	596+
Translucent component	52	306
Blowing		
Coating processes	117	124
Compositions	106	47+
Cutters (see search notes)	83	6
Design	d36	
Apparatus	65	187+
Electric furnaces	13	6
Gratings optical	88	1
Grinding processes	51	283+
Handling cylinders of	214	3.1
Laminated and safety	161	192
Wire glass	161	89+
Laminate, non - structural	161	192+
Processes of uniting glass	65	36
Manufacturing	65	
Filament or fiber making	65	1+
Process	65	17+
With electric lamp making	316	
With metal founding	22	
With metal working	29	
Ornamenting		
By abrasive blasting method	51	317+
By grinding	51	
Plate		
Grinding and polishing	51	
Pot furnaces	263	11+
Pots	263	48
Scriber's	33	18+
Structural	161	
Wire mesh	161	89
Treating hard glass	65	
Apparatus	65	348+
Process	65	111+
GLASSES		
Eye and spectacles	351	41+
Field and opera	88	34+
Ground for cameras	95	49
Sight for liquid level gauge	73	323+
GLAZE		
Compositions	106	48+
GLAZIERS POINTS	85	15
Setters	227	
GLAZING		
Earthenware	117	125
Firing	264	60+
Fruits and vegetables	99	102
GLIDER		
Aerial toy	46	79+
Aircraft	244	16
Porch swing	297	282
Hammock couch type	5	124+
GLIDES	16	42
GLOBE		
Clock	58	44
Lantern	240	100
Design	d48	16
Plants	PLT	89
Operators for	240	30+
Manipulating pole	294	20+
Operator tube lantern	240	30+
Railway amusement	104	68
Teaching	35	46+
GLOVE	2	159
Baseball	2	19
Boxing	2	18
Buttoners	24	40
Forms	223	78+
Knitted	66	174
Circular machine	66	45
Straight machine	66	65
Sewing machine	112	16+
GLOVER'S TOWER	23	283
GLOW		
Discharge diode	313	182+
Transfer counter	315	84.5+

INDEX TO CLASSIFICATION

	Class	Subclass
GLOWER (SEE FILAMENT,LAMP,ELECTRIC, ELECTRODE)		
GLUCAMINES	260	211
GLUCONIC ACID	260	535
GLUCOSE	127	30
Making	127	36+
GLUCOSIDES	260	210+
GLUE (SEE ADHESIVE)	260	117+
Compositions	106	125+
Containing	106	125+
GLUING APPARATUS	118	
Clamps	269	
Pots	126	284
Veneer	156	
Presses	144	281
GLUTAMATES	260	534
Foods ctg	99	16
Preparation by protein hydrolysis	260	529
GLUTATHIONE	260	112
GLYCERIC ACID	260	535
GLYCERIDES (SEE FATS)	260	398+
Alkyd resins	260	75+
Rosin	260	104
GLYCERINE	260	635+
Fermentative production	195	32+
GLYCEROBORATES	260	462
GLYCEROL (SEE GLYCERINE)		
GLYCEROPHOSPHATES	260	461
GLYCIDOL	260	348+
GLYCINE	260	534
GLYCOLS	260	635+
Bori borate	260	462
Ethers	260	615
Oleate	260	410.6
Silicate	260	448.8
Stearate	260	410.6
GLYCOSIDES	260	210
Cardiac	260	210.5
GLYOXAL	260	601
GLYOXALINE	260	309
GLYPTAL T.M.(SEE ALKYD RESINS)		
GOAD	231	2
GOCARTS(SEE BABY CARRIAGE)		
GODET	18	8
Strand feeding wheel	226	168+
GOGGLES	2	14
GOLD		
Alloys	75	165
Beating	72	420+
Carbon compounds	260	430
Electrolysis		
Coating	204	46+
Synthesis	204	109+
Hydrometallurgy	75	118
Laminate	161	213
Metal to metal	29	199
Plate	29	199
Pyrometallurgy	75	83
GOLF		
Bag	150	1.5
On wheeled carrier	280	47.26
Ball	273	62
Making	156	146
Club	273	77+
Miniature course	273	32
Practice devices	273	35+
Score registers	235	
Simulated game	273	87
Tees,holes,greens,devices	273	32+
GONDOLA		
Railway		
Drop bottom	105	244+
Freight	105	406
Inclined bottom	105	256
Side door	105	258+
GONG(SEE BELL)		
GONIOMETER		
Crystal testing	88	14
Light ray type		
Horizontal angle	33	72

FIGURE 6.4
PATENT INDEX SUBCLASSIFICATION

CLASS 2 APPAREL

2-1

JANUARY 1979

1	MISCELLANEOUS
2	GUARDS AND PROTECTORS
2.1 R	.Diving type
2.1 A	...Space suits
2.5	.Penetration resistant
410	.For wearer's head
4	..Insect repelling
5	..Firemen's helmets
6	..Military or aviators' helmets
7	..Heat resistant
8	...High temperature
411	..Including energy-absorbing means
412	...By diverse laminae
413	...By fluid-containing cushion
414	...By interior pads
415	....Including neck pad
416	...By suspension rigging
417	..Including adjustment for wearer's head size
418	...For circumference of crown
419	....And height of crown
420	....By plural-part rigging
421	..Including helmet-retention means
422	..With article-attaching means
423	..And ears
424	..And face
425	..Sport headgear
15	..Eye shields (e. g., hoodwinks or blinds, etc.)
10	...Hat or cap attachments
11	...Hand or body supported
12	...Shades
13	....Spectacle attachments
426	...Goggles
427	....Included in shield for face
428	.....With seal for face
429	......And detachable face plate
430	......And wide-angle lens
431	....Included in shield for eyes
432	.....Having anti-glare shield or lens
433	......By limited-view opening
434	.....Having lens-cover plate
435	.....Having anti-fog shield or lens
436	......By ventilation of shield
437	.......Via tortuous air path
438	.....With movable element external
439	.....Having unitary frame
440	.....With seal (e. g., cup) for each eye
441	......Detachable lens-mounting
442	.....Having frame for each eye
443	......Detachable lens-mounting
444	......Suspended within a goggle
445	.....Connected at center
446	......By nose-rest
447	.....Having wide-angle lens
448	....Including temple element
449	.....And side shield
450	.....And connection to frame
451	....Pivoted side shield
452	....On head band
453	....On horizontal pivot
454	....Foldable or collapsible
9	..Face
16	.Hand or arm
17	..Handle or rein attachments
18	..Boxing gloves
19	..Baseball gloves
20	..Hand pads
21	..Finger cots or protectors
22	.Leg
23	..Trouser attachments
24	..Knee pads or rests
	BODY BRACES AND SUPPORTS
44	.Shoulder and back
45	.Shoulder
46	GARMENT PROTECTORS
47	.Skirts
48	.Aprons

49 R	..Bibs
49 A	....Combined or associated with trays or tray covers
50	..Barbering
51	..Workmen's
52	..Ties and supports
53	.Armpit shields
54	..Combined with garments
55	..Body supported
56	..Dress-attaching features
57	..Frames
58	..Methods of making
59	.Sleeve
60	.Collar or cuff
61	.Stocking
62	.Knee
63	.Try-on hat linings
64	BURIAL GARMENTS
65	FUR GARMENTS
66	.Muffs
67	BATHING GARMENTS
68	.Caps
69	BODY GARMENTS
69.5	.Bag type
	.Union type
70	..Separable
	..Skirted
71	...Combined bifurcated
72	...Convertible bifurcated
73	...Underwear
74	...Dresses
75	...Children's
76	...Waistbands, adjustable or elastic
	..Bifurcated
77	...Men's outer shirts
78 R	...Underwear
78 A	.....Combination nether
78 B	.....Seat and crotch
78 C	.....Elastic insert
78 D	.....Supporting
79	...Trousers and overalls
80	...Children's
81	...Heat resistant
82	...Water resistant
83	...Bed garments
84	.Hooded
85	.Overcoats
86	..Convertible to bifurcated
87	..Waterproof
88	.Capes
89	..Convertible tents
90	.Sweaters
91	.Mufflers
92	.Back and chest protectors
93	.Coats
94	..Hunters' and special-article carrying
95	..Combined with vest or shirt
96	..Front closures
97	..Linings
98	..Collars
99	...Springs
100	...Fasteners
101	..Putting-on accessories
102	.Vests
103	..Combined with shirt or dickey
	.Waists
104	..Nursing
105	..Dress
106	..Blouses
107	...Garment supporting
108	..Jacket type
109	..Underwear
110	...Corset covers
111	...Infants'
112	...Garment supporting
	.Shirts
113	..Undergarments
114	..Bed garments

Another patent office publication, the *Official Gazette: Patents* (C21.5:(vol./no.)), contains summaries and drawings of the approximately one thousand five hundred patents granted each week (figure 6.5). It has been published weekly since 1872. Patent summaries in the *Official Gazette* are arranged in numerical order and are assigned numbers to reflect the order of the U.S. Patent Classification System. Once a searcher is familiar with this classification system, he or she can scan the *Official Gazette* regularly to keep up with the developments in specific technical areas. Each issue of the *Official Gazette* also provides indexes by class and subclass (figure 6.6, page 98), and by name of patentee (figure 6.7, page 99).

The *Index of Patents* (C21.5/2:(yr./part)) is an annual index to information from the *Official Gazette* of the past year. Part I of the *Index of Patents*, the List of Patentees, is an alphabetical list of patentees and patent assignees. For a subject approach, the searcher can turn to part II, Index to Subjects of Invention, which lists all patents assigned during the previous year according to class and subclass.

The *Microfilm List* is another valuable search aid that can be used when a class and subclass have been identified. The base list provides a complete enumeration of patent numbers by class and subclass covering the years 1836 to 1969. To use the *Microfilm List*, first search under class/subclass to find a listing of all patent numbers issued for that class/subclass for the time period covered. Next, turn to the "Cross-Reference" section of the *Microfilm List* to find patent numbers of patents in related classifications.

To bring the search completely up to date, the gap must be bridged between the most recent cumulation of the *Microfilm List* and the present. This can be accomplished by searching each annual *Index of Patents* published after the *Microfilm List* cutoff date. In part II: Subjects of Invention, the index offers a section arranged by class and subclass. To bring the search up to the current week, the searcher must check the weekly *Official Gazette: Patents* published since the last available *Index of Patents*. The *Official Gazette: Patents* can also be searched by class/subclass, allowing the search to identify additional patent numbers issued in that category.

PATENT AND TRADEMARK OFFICE

The Patent and Trademark Office offers the only complete U.S. patent collection in subject arrangement (filed according to the U.S. Patent Classification System). This collection is located in the Public Search Room of the Patent and Trademark Office in Arlington, Virginia. The Search Room is open to the public and allows examination of United States patents granted since 1836. In addition, the Patent and Trademark Office maintains a Scientific Library and a Record Room, both open to the public. The Scientific Library includes scientific and technical books in numerous languages, periodicals focusing on science and technology, the official journals of foreign patent offices, and copies of foreign patents. The Record Room provides for public inspection of records and files on issued patents, as well as other open records.

Although these facilities offer a wealth of patent and scientific information, it is not necessary to visit them in order to complete a patent search. About thirty patent depositories are available in libraries across the country. A list of the

FIGURE 6.5
SAMPLE PAGE FROM THE *OFFICIAL GAZETTE: PATENTS*

iary signal are substantially visually cancelled from the display on said cathode ray tube;
selectively operated switching circuit means responsive to at least some of the synchronizing signal components of the composite television signal for processing said combined signal to cause the redundant auxiliary signal portions to be visually reinforced so that the auxiliary signal components are reproduced on said cathode ray tube; and
control means for selectively initiating operation of said switching circuit means.

4,051,533
SIGNAL PROCESSOR FOR REDUCING INTERFERENCE BETWEEN FREQUENCY-MODULATED SIGNALS
Frank Anthony Griffiths, 105 Hillcroft Crescent, Oxhey, Hertfordshire, England
Filed Sept. 21, 1976, Ser. No. 725,281
Claims priority, application United Kingdom, Sept. 24, 1975, 39195/75
Int. Cl.² H04N *7/06;* H04B *15/00*
U.S. Cl. 358—167 **6 Claims**

1. A method of reducing the interference between a primary frequency-modulated carrier signal and a secondary frequency-modulated carrier signal, comprising both increasing the amplitude of the secondary frequency-modulated carrier signal and decreasing at least those sideband components of the primary carrier signal in the region of the secondary carrier signal as those sideband components tend to increase to interfere with the secondary carrier signal.

4,051,534
HEAD ATTACHED TELEVISION
Peter P. Dukich, Blaine; Isaac W. Metzger, Robbinsdale, and John A. Volk, West St. Paul, all of Minn., assignors to Honeywell Inc., Minneapolis, Minn.
Filed Oct. 27, 1976, Ser. No. 735,934
Int. Cl.² H04N *5/30*
U.S. Cl. 358—210 **4 Claims**

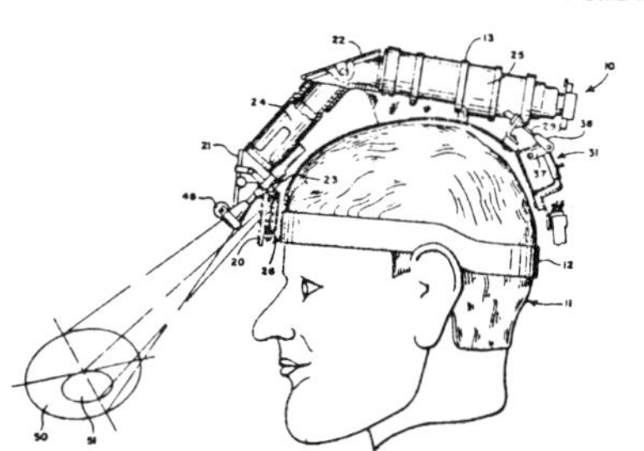

1. In a remotely controlled head-mounted visual apparatus for viewing manual operations carried on by the wearer, said apparatus including camera means for generating signals representative of a visual image received, aiming and focusing means for said camera means, remote control means for controlling said aiming and focusing means, means for reconverting said signals to reproduce said visual image, means for transmitting said signals from said camera means to said means for reconverting, real-time viewing means for viewing said reconverted image, and mounting means for mounting said camera, aiming and focusing means on the head of said wearer, the improvement comprising:
forward directed illumination means fixed on said head-mounted apparatus in a manner such that the location of the field lighted thereby is determined locally by the attitude of the head of said wearer.

4,051,535
MAGNIFICATION OF TELEVISION IMAGES
James M. Inglis, 24 Cotton Ave., Braintree, Mass. 02184
Filed Apr. 9, 1976, Ser. No. 675,603
Int. Cl.² H04N *5/72*
U.S. Cl. 358—231 **10 Claims**

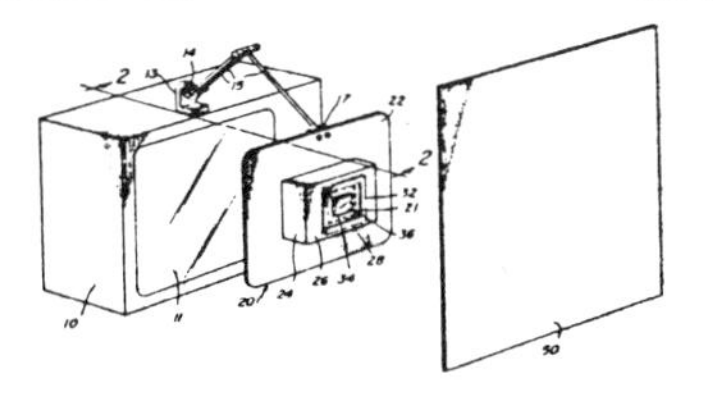

1. A system for magnifying a television image comprising:
a fresnel lens of focal length up to 10 inches,
an enclosure fitting all around the circumference of said lens and extending in a direction parallel to the optical axis of the lens,
a means defining an aperture in the end of said enclosure away from said lens,
a means for intercepting light arranged in combination with said lens so that if a television image source is placed on the side of the lens opposite to the direction in which said enclosure extends, substantially along the optical axis of the lens and within 18 inches of the lens, substantially all light emitted from said source and not passing through said lens will be intercepted,
the distance between said lens and said aperture being between 20% and 50% of said focal length, and
the area of said aperture being between 25% and 50% of the area of said lens.

4,051,536
ELECTRONIC HALFTONE IMAGING SYSTEM
Paul G. Roetling, Ontario, N.Y., assignor to Xerox Corporation, Stamford, Conn.
Filed Mar. 14, 1975, Ser. No. 558,594
Int. Cl.² H04N *1/22*
U.S. Cl. 358—298 **12 Claims**

1. A method for creating on a light sensitive recording medium a halftone reproduction of an original image comprising the steps of:
a. providing an electronic signal representing the average gray scale of said original image over a halftone dot period;
b. providing a plurality of electronic signals representing details of said original image over said halftone dot period;
c. providing a halftone screen function in electronic form;
d. combining said halftone screen signals of (c) with said original image signals of (b) to provide a plurality of sum functions; and

FIGURE 6.6
SAMPLE CLASSIFICATION OF PATENTS

CLASSIFICATION OF PATENTS

ISSUED DECEMBER 21, 1982

NOTE.—First number, class; second number, subclass; third number, patent number

Subclass	Patent number
CLASS 2	
209.1	4,364,123
234	4,364,124
255	4,364,125
CLASS 3	
1.5	4,364,126
	4,364,127
6.1	4,364,128
CLASS 4	
324	4,364,129
449	4,364,130
495	4,364,131
546	4,364,132
617	4,364,133
CLASS 5	
201	4,364,134
443	4,364,135
CLASS 8	
634	4,364,738
654	4,364,739
CLASS 10	
86 A	4,364,136
CLASS 14	
71.3	4,364,137
CLASS 15	
50 R	4,364,138
104 06 R	4,364,141
104.3 SN	4,364,139
	4,364,140
117	4,364,142
145	4,364,143
229 AC	4,364,144
236 R	4,364,145
323	4,364,146
404	4,364,147
CLASS 16	
32	4,364,148
87 R	4,364,149
126	4,364,150
CLASS 19	
0.2	4,364,151
105	4,364,152
107	4,364,153
129 R	4,364,154
CLASS 24	
217 R	4,364,155
CLASS 26	
2 R	4,364,156
CLASS 28	
179	4,364,157
CLASS 29	
116 AD	4,364,158
156.5 R	4,364,159
156.8 H	4,364,160
407	4,364,161
570	4,364,163
571	4,364,164
	4,364,165
	4,364,166
576 B	4,364,167
592 R	4,364,168
596	4,364,169
724	4,364,170
828	4,364,171
858	4,364,172
882	4,364,173
CLASS 30	
153	4,364,174
CLASS 33	
1 H	4,364,175
18 R	4,364,176
147 K	4,364,177
174 H	4,364,182
174 L	4,364,178
	4,364,179
	4,364,180
	4,364,181
268	4,364,183
447	4,364,184

Subclass	Patent number
CLASS 34	
122	4,364,185
CLASS 36	
3 B	4,364,186
15	4,364,187
31	4,364,188
	4,364,189
32 R	4,364,190
CLASS 37	
2 R	4,364,191
CLASS 40	
155	4,364,192
CLASS 43	
1	4,364,193
131	4,364,194
CLASS 44	
1 C	4,364,740
51	4,364,741
	4,364,742
66	4,364,743
CLASS 46	
135 R	4,364,195
242	4,364,196
CLASS 47	
56	4,364,197
CLASS 48	
86 R	4,364,744
209	4,364,745
CLASS 49	
62	4,364,198
181	4,364,199
192	4,364,200
248	4,364,201
352	4,364,202
409	4,364,203
CLASS 51	
52 R	4,364,204
298	4,364,746
CLASS 52	
4	4,364,205
79.7	4,364,206
81	4,364,207
82	4,364,208
208	4,364,209
221	4,364,210
245	4,364,211
281	4,364,212
309.17	4,364,213
311	4,364,214
488	4,364,215
731	4,364,216
CLASS 53	
58	4,364,217
331.5	4,364,218
374	4,364,219
411	4,364,220
CLASS 55	
25	4,364,747
27	4,364,748
73	4,364,749
89	4,364,750
96	4,364,751
138	4,364,752
179	4,364,753
269	4,364,754
290	4,364,755
316	4,364,756
357	4,364,757
365	4,364,758
487	4,364,759
523	4,364,760
	4,364,761
CLASS 56	
13.6	4,364,221
328 R	4,364,222
CLASS 57	
5	4,364,223
263	4,364,224
291	4,364,225

Subclass	Patent number
CLASS 60	
276	4,364,226
	4,364,227
398	4,364,228
414	4,364,229
444	4,364,230
577	4,364,231
641.2	4,364,232
712	4,364,233
CLASS 62	
3	4,364,234
55.5	4,364,235
77	4,364,236
99	4,364,242
160	4,364,237
217	4,364,238
235.1	4,364,239
476	4,364,240
505	4,364,241
CLASS 65	
2	4,364,762
22	4,364,763
29	4,364,764
106	4,364,765
160	4,364,766
CLASS 66	
9 B	4,364,243
75.2	4,364,244
78	4,364,245
145 R	4,364,246
163	4,364,247
CLASS 68	
5 E	4,364,248
CLASS 70	
264	4,364,249
456 R	4,364,250
CLASS 71	
86	4,364,767
88	4,364,768
90	4,364,769
CLASS 72	
58	4,364,251
59	4,364,252
187	4,364,253
319	4,364,254
345	4,364,255
356	4,364,256
405	4,364,257
453.15	4,364,258
CLASS 73	
35	4,364,259
	4,364,260
40	4,364,261
53	4,364,262
61.1 C	4,364,263
105	4,364,264
113	4,364,265
115	4,364,266
146	4,364,267
178 R	4,364,268
223	4,364,269
298	4,364,270
432 R	4,364,271
	4,364,272
614	4,364,273
615	4,364,274
649	4,364,275
721	4,364,276
862.34	4,364,277
862.36	4,364,278
862.66	4,364,279
	4,364,280
CLASS 74	
55	4,364,281
424.8 A	4,364,282
489	4,364,283
540	4,364,284
606 R	4,364,285
768	4,364,286
CLASS 75	
53	4,364,770
58	4,364,771
126 R	4,364,772

Subclass	Patent number
CLASS 76	
101 D	4,364,287
CLASS 81	
3 R	4,364,288
9.51	4,364,289
CLASS 82	
36 R	4,364,290
CLASS 83	
100	4,364,291
605	4,364,292
674	4,364,293
796	4,364,294
CLASS 84	
1.15	4,364,295
1.26	4,364,296
440	4,364,297
465	4,364,298
478	4,364,299
CLASS 89	
36 A	4,364,300
CLASS 91	
20	4,364,301
29	4,364,302
329	4,364,303
420	4,364,304
CLASS 92	
63	4,364,305
71	4,364,306
157	4,364,307
CLASS 99	
351	4,364,308
352	4,364,309
357	4,364,310
CLASS 100	
6	4,364,311
CLASS 101	
110	4,364,312
401.1	4,364,313
CLASS 104	
225	4,364,314
CLASS 105	
215 C	4,364,315
CLASS 106	
281 R	4,364,773
287.13	4,364,774
CLASS 111	
3	4,364,316
CLASS 112	
132	4,364,317
147	4,364,318
238	4,364,319
262.2	4,364,320
311	4,364,321
CLASS 114	
71	4,364,322
265	4,364,323
280	4,364,324
331	4,364,325
369	4,364,326
CLASS 118	
102	4,364,327
411	4,364,328
652	4,364,329
697	4,364,330
CLASS 119	
28	4,364,331
48	4,364,332
52 AF	4,364,334
52 R	4,364,333
61	4,364,335
CLASS 123	
1 A	4,364,336
3	4,364,337
41.10	4,364,338
41.42	4,364,339
55 R	4,364,340

Subclass	Patent number
90 17	4,364,341
143 B	4,364,342
179 B	4,364,343
179 BG	4,364,344
198 F	4,364,345
323	4,364,346
339	4,364,347
	4,364,348
	4,364,349
	4,364,350
357	4,364,351
376	4,364,352
425	4,364,353
437	4,364,354
438	4,364,355
440	4,364,356
	4,364,357
	4,364,358
	4,364,359
450	4,364,360
453	4,364,361
454	4,364,362
492	4,364,363
527	4,364,364
557	4,364,365
564	4,364,366
	4,364,367
568	4,364,368
569	4,364,369
575	4,364,370
CLASS 124	
5	4,364,371
CLASS 126	
39 R	4,364,372
418	4,364,373
442	4,364,374
444	4,364,375
CLASS 128	
1.1	4,364,376
1.5	4,364,377
24.5	4,364,378
79	4,364,379
89 A	4,364,380
92 D	4,364,382
92 E	4,364,381
204 28	4,364,384
213 R	4,364,385
214 C	4,364,387
214 E	4,364,386
214 R	4,364,383
234	4,364,388
303 R	4,364,389
303.1	4,364,390
305.3	4,364,391
325	4,364,392
335.5	4,364,393
419 PT	4,364,396
710	4,364,397
736	4,364,398
774	4,364,399
CLASS 131	
277	4,364,400
297	4,364,401
311	4,364,402
332	4,364,403
CLASS 133	
4 A	4,364,404
CLASS 134	
3	4,364,775
10	4,364,776
29	4,364,777
CLASS 135	
70	4,364,405
CLASS 136	
206	4,365,106
258	4,365,107
CLASS 137	
15	4,364,406
71	4,364,407
107	4,364,408
486	4,364,409
489	4,364,410
513.5	4,364,411
557	4,364,412
624.14	4,364,414
624.2	4,364,413

Subclass	Patent number
625.3	Re 31,105
625.32	4,364,415
CLASS 138	
30	4,364,416
42	4,364,417
103	4,364,418
167	4,364,419
CLASS 139	
91	4,364,420
383 A	4,364,421
CLASS 141	
1.1	4,364,422
CLASS 144	
366	4,364,423
CLASS 148	
1.5	4,364,778
	4,364,779
6	4,364,780
6.3	4,364,781
CLASS 149	
21	4,364,782
CLASS 150	
1	4,364,424
51	4,364,425
CLASS 152	
209 R	4,364,426
349	4,364,427
CLASS 156	
69	4,364,783
78	4,364,784
88	4,364,785
99	4,364,786
164	4,364,787
179	4,364,788
214	4,364,789
346	4,364,790
506	4,364,791
628	4,364,792
643	4,364,793
CLASS 159	
17 P	4,364,794
CLASS 162	
158	4,364,795
CLASS 165	
11 R	4,364,428
28	4,364,429
CLASS 166	
214	4,364,430
275	4,364,431
290	4,364,432
339	4,364,433
CLASS 171	
63	4,364,434
CLASS 172	
15	4,364,435
33	4,364,436
349	4,364,437
789	4,364,438
821	4,364,439
CLASS 173	
93.7	4,364,440
CLASS 174	
50	4,365,108
109	4,365,109
CLASS 175	
84	4,364,441
CLASS 177	
177	4,364,44
CLASS 178	
22.10	4,365,11
22.13	4,365,11
CLASS 179	
1 E	4,365,11
1 SC	4,365,11

FIGURE 6.7
SAMPLE LIST OF PATENTEES BY NAME

LIST OF PATENTEES

TO WHOM

PATENTS WERE ISSUED ON THE 23RD DAY OF FEBRUARY, 1982

NOTE—Arranged in accordance with the first significant character or word of the name (in accordance with city and telephone directory practice).

A/S Akers Mek. Verksted: *See—*
Hovind, Leif; and Fladby, Tron-Halvard, 4,316,725, Cl. 55-41.000.
Abbott, Bernard J.; and Berry, Dennis R., to Eli Lilly and Company. Enzymatic deesterification of cephalosporin methyl esters. 4,316,955, Cl. 435-47.000.
Abbott Laboratories: *See—*
Genese, Joseph N.; and Muetterties, Andrew J., 4,316,460, Cl. 128-214.00R.
Nara, Takashi; Okachi, Ryo; Kawamoto, Isao; Sato, Tomoyasu; and Oka, Tetsuo, 4,316,957, Cl. 435-119.000.
Abe, Hiroshi: *See—*
Kawai, Mituo; Fujita, Takashi; Shirai, Hideo; Nakagawa, Masatoshi; and Abe, Hiroshi, 4,316,743, Cl. 75-124.000.
Acebo, William F.: *See—*
Wolf, Irving W.; Stafford, Michael K.; Kahan, Hillard M.; Acebo, William F.; Scott, Lawrence M.; and Lee, Yu C., 4,316,738, Cl. 75-0.5BA.
Acker, William F., to Honeywell Inc. Signal monitor system. 4,317,080, Cl. 328-151.000.
Ackerman, Morris, to Singer Company, The. Maintenance training device. 4,316,720, Cl. 434-224.000.
Adamek, Radomil: *See—*
Kupf, Lubomir; Adamek, Radomil; and Mursec, Mirko, 4,316,803, Cl. 210-455.000.
Adamer, Siegfried, to Societe d'Assistance Technique pour Produits Nestle S.A. Treatment of a coffee extract. 4,316,916, Cl. 426-329.000.
Adams, Joan M.; Shoaf, Myron B.; Bochmann, Carl E.; and Basile, Peter A., to General Foods Corporation. Carbonated beverage container. 4,316,409, Cl. 99-275.000.
Adams, John T.; Kompelien, Arlon D.; Nelson, Marvin D.; and Pinckaers, B. Hubert, to Honeywell Inc. Energy saving thermostat. 4,316,577, Cl. 236-46.00R.
Adell, Robert, to U.S. Product Development Co. Door edge guard. 4,316,348, Cl. 49-462.000.
Adria Laboratories, Inc.: *See—*
Alam, Abu S.; and Eichel, Herman J., 4,316,884, Cl. 424-19.000.
Aesculap-Werke Aktiengesellschaft, vormals Jetter & Scheerer: *See—*
Braun, Karl; and Wintermantel, Erich, 4,316,470, Cl. 128-346.000.
Agence Nationale de Valorisation de la Recherche: *See—*
Charpak, Georges, 4,317,038, Cl. 250-385.000.
Agence Nationale de Valorisation de la Recherche (ANVAR): *See—*
Atlani, Martial; Loutaty, Robert; Wakselman, Claude; and Yacono, Charles, 4,316,796, Cl. 208-313.000.
Fruman, Daniel; and Tulin, Marshall, 4,316,383, Cl. 73-55.000.
Agency of Industrial Science and Technology, The: *See—*
Kudo, Bosshi; and Yoshioka, Masamichi, 4,316,764, Cl. 156-617.00H.
AGFA-Gevaert Aktiengesellschaft: *See—*
Gernert, Herbert, 4,316,953, Cl. 430-569.000.
Agri-Canvas Inc.: *See—*
Verbeek, John M., 4,316,536, Cl. 198-699.000.
Aiello, Samuel, Jr., to Ilo Engineering, Inc. Combined battery holder and switch. 4,317,161, Cl. 362-103.000.
Air Factors West: *See—*
Lambert, Robert R., 4,316,407, Cl. 98-40.00D.
Air Products and Chemicals, Inc.: *See—*
Ford, Michael E.; and Johnson, Thomas A., 4,316,840, Cl. 260-239.0BC.
Ford, Michael E.; and Johnson, Thomas A., 4,316,841, Cl. 260-239.0BC.
Aktiebolaget Overums Bruk: *See—*
Lindqvist, Rolf E., 4,316,507, Cl. 172-225.000.
Akune, Mikio; and Kinoshita, Yoshiaki, to Nittetu Chemical Engineering Ltd. Method for the combustive treatment of waste fluids containing nitrogen compounds. 4,316,878, Cl. 423-235.000.
Akutsu, Yoji: *See—*
Kita, Hisanao; Karatsu, Yoshinori; Nakazaki, Takamitsu; and Akutsu, Yoji, 4,317,022, Cl. 219-121.0EU.
Alam, Abu S.; and Eichel, Herman J., to Adria Laboratories, Inc. Sustained release pharmaceutical formulation. 4,316,884, Cl. 424-19.000.
Alan I. Gerald Corporation: *See—*
Kahn, Harvey, 4,316,340, Cl. 42-66.000.
Albenberger, Johann: *See—*
Lenz, Wolfgang; Albenberger, Johann; and Knell, Karl, 4,316,631, Cl. 297-284.000.
Alberts, Steven L., to GTE Products Corporation. Voltage monitoring and indicating circuit. 4,317,056, Cl. 307-350.000.
Albright, Jane: *See—*
Gerlock, John L.; Braslaw, Jacob; and Albright, Jane, 4,316,992, Cl. 568-621.000.
Aleinikov, Nikolai A.: *See—*
Zelenov, Petr I.; Usachev, Petr A.; Davydov, Jury V.; Lyakhov, Vyacheslav P.; Zelenova, Irina M.; Aleinikov, Nikolai A.; Sladkovich, Vladlen F.; and Titov, Viktor I., 4,316,542, Cl. 209-39.000.
Alexeev, Boris N. Artificial crystalline lens. 4,316,292, Cl. 3-13.000.
Allen, Charles B. Eyeglass frame with pocket clip. 4,316,654, Cl. 351-155.000.
Allen Engineering Corporation: *See—*
Allen, J. Dewayne, 4,316,715, Cl. 425-456.000.
Allen, J. Dewayne, to Allen Engineering Corporation. Vibratory concrete screed having an adjustable extension bracket. 4,316,715, Cl. 425-456.000.
Allen, William P., Jr.; and Zieg, Benjamin S., to Lockheed Corporation. Conformal HF loop antenna. 4,317,121, Cl. 343-712.000.
Allied Corporation: *See—*
DeCristofaro, Nicholas J.; and Henschel, Claude, 4,316,573, Cl. 228-263.00R.
Sexton, Peter; and DeCristofaro, Nicholas J., 4,316,572, Cl. 228-263.00R.
Tunick, Allen A.; Largman, Theodore; and Sifniades, Stylianos, 4,316,877, Cl. 423-10.000.
Vermeer, Dick C.; and Biron, Raymond J., 4,316,312, Cl. 28-255.000.
Alstad, Sven O.: *See—*
Petersson, Stefan; and Alstad, Sven O., 4,316,709, Cl. 425-174.80E.
Altenschulte, Raymond A., to RCA Corporation. A.C. Power line assembly. 4,317,152, Cl. 361-104.000.
Alvarez, Luis W.; and Schwemin, Arnold J., to Schwem Instruments. Stabilized zoom binocular. 4,316,649, Cl. 350-16.000.
Alvarez, Luis W. Stand alone collision avoidance system. 4,317,119, Cl. 343-112.0CA.
Amacker, Inc.: *See—*
Amacker, Joseph A., 4,316,526, Cl. 182-135.000.
Amacker, Joseph A., to Amacker, Inc. Apparatus for and method of climbing an upright columnar member. 4,316,526, Cl. 182-135.000.
Ambrose, Richard J.: *See—*
Hergenrother, William L.; Schwarz, Richard A.; Ambrose, Richard J.; and Hayes, Robert A., 4,316,967, Cl. 525-111.000.
Amerace Corporation: *See—*
Siebens, Larry N., 4,316,646, Cl. 339-7.000.
American Cyanamid Company: *See—*
Pfeiffer, Ronald E.; and DeMaria, Francesco, 4,316,714, Cl. 425-382.200.
American Electronics, Inc.: *See—*
Gunderson, Norman R., 4,316,655, Cl. 352-166.000.
American Hospital Supply Corporation: *See—*
Stoneback, W. Keith, 4,316,455, Cl. 128-132.00D.
Stoneback, W. Keith, 4,316,456, Cl. 128-132.00D.
American Standard Inc.: *See—*
Bramwell, John S., 4,316,864, Cl. 264-86.000.
American Sterilizer Company: *See—*
Hopper, James A., 4,316,726, Cl. 55-89.000.
American Sussen Corp.: *See—*
Steiner, Erwin, 4,316,370, Cl. 68-5.00D.
AMF Incorporated: *See—*
Cummins, Donald L., 4,316,534, Cl. 198-345.000.
Lundberg, Duane R., 4,316,608, Cl. 272-117.000.
AMP Incorporated: *See—*
Mosser, Benjamin H., III, 4,317,006, Cl. 178-46.000.
Ampex Corporation: *See—*
Monforte, Frank R.; and Argentina, Giltan M., 4,316,923, Cl. 428-68.000.
Wolf, Irving W.; Stafford, Michael K.; Kahan, Hillard M.; Acebo, William F.; Scott, Lawrence M.; and Lee, Yu C., 4,316,738, Cl. 75-0.5BA.
Anantha, Narasipur G.; and Chang, Augustine W., to International Business Machines Corporation. Method for making a high sheet resistance structure for high density integrated circuits. 4,316,319, Cl. 29-577.00C.
Andersen, Dennis H., to MTS Systems Corporation. Deep drawing press with blanking and draw pad pressure control. 4,316,379, Cl. 72-351.000.
Andersen, Helge H., to Kongskilde Koncernselskab A/S. Implement frame, especially for agricultural machines. 4,316,511, Cl. 172-776.000.
Anthony Manufacturing Company: *See—*
Ray, Charles A.; and Kent, John L., 4,316,579, Cl. 239-123.000.
Antoshkiw, Thomas; Cannalonga, Marco A.; and Guerin, Frank, to Hoffman-La Roche Inc. Stable carotenoid solutions. 4,316,917, Cl. 426-540.000.

patent depository libraries follows. These depositories have copies on file, usually arranged in numerical order by patent number, and can offer assistance in beginning a search. Many other non-depository libraries have basic patent search tools in their collections, enabling a preliminary search to be performed. In addition, patents can be computer-searched using numerous government machine-readable data files, discussed later in this chapter.

Patent Depository Libraries

Albany, NY—State University of New York Library
Atlanta, GA—Georgia Institute of Technology Library
Baton Rouge, LA—Louisiana State University
Birmingham, AL—Public Library
Boston, MA—Public Library
Buffalo, NY—Buffalo & Erie County Public Library
Charleston, SC—Medical University of South Carolina
Chicago, IL—Public Library
Cincinnati, OH—Public Library
Cleveland, OH—Public Library
Columbus, OH—Ohio State University Library
Dallas, TX—Public Library
Denver, CO—Public Library
Detroit, MI—Public Library
Durham, NH—University of New Hampshire
Houston, TX—Rice University, Fondren Library
Kansas City, MO—Linda Hall Library
Lincoln, NE—University of Nebraska, Love Library
Los Angeles, CA—Public Library
Madison, WI—Engineering & Physical Sciences Library, University of Wisconsin
Memphis, TN—Memphis and Shelby County Public Library
Milwaukee, WI—Public Library
Minneapolis, MN—Public Library
Newark, DE—University of Delaware
Newark, NJ—Public Library
New York, NY—Public Library
Philadelphia, PA—Public Library
Pittsburgh, PA—Carnegie Library
Providence, RI—Public Library
Raleigh, NC—North Carolina State University, D. H. Hill Library
Sacramento, CA—California State Library
Seattle, WA—University of Washington Engineering Library
St. Louis, MO—Public Library
Stillwater, OK—Oklahoma State University Library
Sunnyvale, CA—Sunnyvale Patent Clearinghouse*
Tempe, AZ—Arizona State University, Science Library
Toledo, OH—Public Library
University Park, PA—Pennsylvania State University Libraries

*Collection arranged by subject matter.

OTHER PRINT SOURCES

The major indexes of the National Technical Information Service, Department of Defense, and National Aeronautics and Space Administration (*Government Reports Announcements and Index, Technical Abstract Bulletin*, and *Scientific and Technical Aerospace Reports*, respectively) each list government-owned patents. NTIS and NASA publish additional sources for patent bibliographical control, to supplement their print indexes and computer data files. In addition, a number of government-produced indexes and abstracts cover patents along with technical reports, journal articles, and other formats. These sources are described below.

The National Technical Information Service is the central clearinghouse for information concerning new United States government-owned patents and patent applications. This information is summarized in the NTIS weekly abstract newsletter, *Government Inventions for Licensing*, while selected commercially promising inventions are summarized in the monthly NTIS *Tech Notes* publication, *Selected Technology for Licensing*.

NTIS has also compiled a *Catalog of Government Patents*, which offers information on thousands of government-owned inventions, many of which are available for licensing. Part A of the *Catalog* provides a listing of patents with abstracts and in some cases, illustrations, arranged according to the Patent and Trademark Office Classification System. Part A covers patents issued to federal agencies between 1966 and 1974; part B covers 1975 through 1980. Part B lists patents according to a unique subject arrangement that incorporates the Patent and Trademark Office Classification System in addition to special subcategories developed by NTIS. Inventions are indexed by both subject and agency in part B. Part C of the *Catalog* is the weekly newsletter *Government Inventions for Licensing*.

Each year NASA patents many of the inventions that have resulted from NASA-supported research. These NASA-owned patents and patent applications are listed in the semiannual *NASA Patent Abstracts Bibliography* (*NASA PAB*). Each issue of this publication has separately bound abstract and index sections and includes citations which were originally published in NASA's *Scientific and Technical Aerospace Reports* (*STAR*) since May 1969.

NASA PAB offers five indexes which are cross-indexed: subject index, inventor index, source index (name of inventing organization), number index (listing NASA case number, patent application number, U.S. Patent Classification number, U.S. Patent number, NASA subject category number, and NASA accession number), and accession number index.

COMPUTER SEARCHING

In addition to print resources for information on patents, there are several U.S. government machine-readable data bases that provide computer searches of patent data files. Most of these include patents as one of many types of documents indexed. In addition, many of these are discipline oriented, listing only those patents that fall into their specific subject coverage.

Chart 6.1 (pages 102-105) lists several government-produced data bases which include patents in their coverage. Three which are interdisciplinary in

CHART 6.1
GOVERNMENT DATA BASES CONTAINING PATENT INFORMATION

Data Base Name	Producer
Air Pollution Technical Information Center	EPA
DOE Energy Data Base	DOE
Defense Technical Information Center Data Base	DoD
NASA STI Database	NASA
NTIS Bibliographic Data File	NTIS
Nuclear Safety Information Center	Oak Ridge National Laboratory

CHART 6.1 (cont'd)

Coverage
Worldwide coverage of literature on air quality and air pollution prevention and control.
International literature on various energy fields.
Classified technical reports resulting from U. S. Government-sponsored research.
Worldwide technical report literature related to aerospace information.
Unclassified technical reports resulting from U. S. Government-sponsored research.
Literature related to nuclear power plants and research reactors.

(Chart 6.1 continues on page 104).

CHART 6.1 (cont'd)

Date Base Name	Producer
Nuclear Structure References	DOE
Office of Technology Assessment and Forecast Data Base	OTAF, U. S. Patent and Trademark Office
Scientific and Technical Aerospace Reports (STAR)	NASA
Selected Water Resources Abstracts	Dept. of the Interior
Solid Waste Information Retrieval System	EPA

CHART 6.1 (cont'd)

Coverage
Nuclear physics literature.
Covers all U. S. patents since 1790.
Technical reports related to aerospace. A subset of the NASA STI Database.
Water-related aspects of the life, physical and social sciences; engineering and legal aspects of water use, control, management, conservation, and characteristics.
International literature on solid waste management.

scope are the Defense Technical Information Center Data Base, the NTIS Bibliographic Data File, and the Office of Technology Assessment and Forecast Data Base. Both the Defense Technical Information Center and NTIS data bases cover U.S. government-owned patents only. The Office of Technology Assessment and Forecast Data Base of the U.S. Patent and Trademark Office covers all U.S. patents, from 1790 to the present. These three data bases will be discussed in some detail below.

The Office of Technology Assessment and Forecast (OTAF) was established within the Patent and Trademark Office in 1971. Its mission is to encourage and enhance the use of the Patent and Trademark Office patent file. This mission involves assembling, analyzing, and providing access to meaningful data about the file. To this end, the Office of Technology Assessment and Forecast has constructed a master data base of all U.S. patents.

This data base has been used by OTAF to perform two functions. One use is in compiling patent-related publications. Publications in the Technology Assessment and Forecast Reports series have provided summaries of patent activity by U.S. and foreign residents, reviews of highly active technological areas, and comparisons between patent and economic activity in specific Standard Industrial Classification categories. A second publication series, Patent Profiles, focuses on patent activity in selected technologies, such as synthetic fuels, solar energy, and microelectronics.

The OTAF data base is also employed to prepare custom reports, tailored to satisfy individual or agency requests for patent data. Data elements in the OTAF data base can be manipulated to satisfy specific information requests, with several format options such as charts, tables, or graphs. This service is provided on a cost reimbursable basis, for which an estimate can be obtained prior to contracting the service. Searches of the OTAF data base are available only through OTAF (contact Custom Reports Manager, OTAF, U.S. Patent and Trademark Office, Washington, DC 20231, [703]557-2982).

CASSIS is a search service sponsored by the U.S. Patent and Trademark Office that gives patent depository libraries direct, online access to Patent and Trademark Office data. Currently available at patent depository libraries, CASSIS has been designed to help users of these patent collections.

The system offers searches by patent numbers, classifications, and keywords in classification titles. Searches online on CASSIS may eliminate the need to search manually, or can help in defining a "field of search" and in identifying patents in that field.

Both the Defense Technical Information Center and the National Technical Information Service (NTIS) serve as clearinghouses for documentation on government-owned patents and patent applications. The focus of the NTIS data base is on unclassified information and is thus open to the general public; the Defense Technical Information Center offers restricted access to classified information.

As the patent assignee for inventors engaged in Department of Defense sponsored research, the DoD holds patent rights to numerous inventions. Access to information on these inventions is available to members of the Department of Defense, other government agencies, and their contractors through the Defense Technical Information Center, and to the general public through NTIS.

DoD registered users can request searches of the Defense Technical Information Center data base to access the data file on DoD-sponsored patents and patent applications. Members of the general public can access DoD patent records through the print and machine-readable resources of NTIS, including the *Catalog of Government Patents, Government Inventions for Licensing, Government Reports Announcements and Index*, and the NTIS data base. In addition to DoD-owned patents, the NTIS data base also offers documentation on other government-owned patents resulting from government-sponsored research.

NTIS's Patent Data Base is produced by the Patent and Trademark Office, and has three files. The bibliographic file, covering 1970 to the present, contains bibliographic information about each patent that appears in the *Official Gazette*. The file does not store drawings, but does include the abstract, patent title, inventor's name, date of issuance, filing date, assignee, and other information. A second file, the full text file, covers the same time period and contains the full text of each patent exclusive of drawings. The classification file contains classification information on all patents ever issued by the Patent and Trademark Office.

Other patent files available from NTIS are

The Concordance File—a guide to relating the U.S. Patent Classification System to the International Patent Classification.

The Index to Classification—listing patents alphabetically by subject headings assigned to classes and subclasses. This file is intended to act as an initial entry into the classification system, and is particularly useful to individuals unfamiliar with the classification system or the technology under study.

Technology Subjects Files—provide technological information on patents issued under class 364 subclass 200, "General Purpose Digital Processing Systems," and under class 364 subclass 900, "Special Purpose Digital Processing Systems."

Sequential Classification Title File (Manual of Classification)—a comprehensive listing of the U.S. Patent Classification System, the *Manual of Classification* in a machine-readable format.

For information on the availability of any of these files, contact NTIS, 5285 Port Royal Road, Springfield, VA 22161; (703)487-4807.

COMMERCIAL PATENT DATA BASES

There are several commercially produced patent data bases. Pergamon International Information Corporation, for example, produces PATSEARCH and PATLAW. PATSEARCH is a comprehensive data file of patent front-page information, including abstracts. PATLAW retrieves complete headnotes and other bibliographic information from cases reported since 1967 in *United States Patents Quarterly*.

Other commercial patent search systems may be identified by using directories of data bases. Several of these are listed in chapter 11, Data Bases, in this book. Patents can also be retrieved using non-governmental indexing and abstracting services such as Chemical Abstracts and World Aluminum Abstracts.

FOREIGN PATENTS

Approximately one hundred twenty countries offer protection for inventions, and both patent laws and practice differ from country to country. Individuals seeking patent protection in other nations must apply under those nations' laws and procedures. Since the rights granted by a United States patent do not apply in foreign countries, and vice versa, inventors often seek patent protection in more than one country. Because it is an important market, significant foreign inventions are frequently patented in the United States in addition to the inventor's own country. A result is that U.S. patent documentation reflects not only U.S. technical advances, but international technological efforts as well. In 1980, for example, about 40 percent of all U.S. patents issued were granted to citizens of foreign countries.

FOREIGN PATENT SEARCHING

When searching foreign patents it is helpful to realize that most countries publish some form of official patent journal, similar in intent to the *Official Gazette* in the United States. Individuals wishing to search foreign patent literature may wish to refer to the U.S. Patent and Trademark Office collection of official journals issued by foreign patent offices, as well as their files of foreign patents.

A helpful guide for foreign patent searching is Kase's *Foreign Patents*, which concentrates on official sources of foreign patent literature and foreign specifications and applications.[5] Another useful source is the *Concordance: United States Patent Classification to International Patent Classification*, a guide for relating the U.S. Patent Classification System to the third edition of the International Patent classification (IPC).[6]

The International Patent Documentation Center (INPADOC) is a central clearinghouse for patents worldwide. INPADOC, located in Vienna, publishes *The INPADOC Patent Gazette* and maintains a centralized data base which includes patent documents from 46 countries, and covers 95 percent of patents in the world. Another foreign patent data base is the World Patents Index from Derwent Publications, Ltd. of London. INPADOC is available for online searching on DIALOG; WPI is available on ORBIT.

OBTAINING PATENT COPIES

Printed copies of any patent, identified by its patent number, may be purchased from the Commissioner of Patents, Patent Office, U.S. Department of Commerce, Washington, DC 20231. Prices are as follows: Patents (except design patents and color plant patents)—$.50; plant patents (in color)—$1.00; design patents—$.20. Commercial organizations such as IFI/Plenum Data Company (2001 Jefferson Davis Highway, Arlington, VA 22202) and Research Publications, Inc. (2221 Jefferson Davis Parkway, Arlington, VA 22202) can be contacted for obtaining copies of patents. You may also write Public Search Room, Crystal Plaza, 2021 Jefferson Davis Highway, Arlington, VA or OTAF Computer Searches, Gary Robinson, Custom Reports Manager, Office of

Technology Assessment and Forecast, U.S. Patent and Trademark Office, Washington, DC 20231; (703)557-2982.

REFERENCES

1. John T. Maynard, "How to Read a Patent," *IEEE Transactions on Professional Communication* PC-22 (June 1979): 112.
2. Ibid., p. 112.
3. Ibid., p. 112.
4. Marshall C. Dann; Herbert C. Wamsley; and Barry L. Grossman, "Patent and Trademark Office, U.S.," in Allen Kent, Harold Lancour, and Jay E. Daily, eds., *Encyclopedia of Library and Information Science*, Vol. 21 (New York: Marcel Dekker, 1977), p. 448.
5. Francis J. Kase, *Foreign Patents: A Guide to Official Patent Literature* (Dobbs Ferry, NY: Oceana Publications, Inc., 1972).
6. U.S. Department of Commerce, Patent and Trademark Office, *Concordance: United States Patent Classification to International Patent Classification* (Washington, DC: U.S. Government Printing Office, 1980) (C21.14/2:C74/980).

FURTHER READING

Dann, Marshall C.; Herbert C. Wamsley; and Barry L. Grossman. "Patent and Trademark Office, U.S.." In Allen Kent, Harold Lancour, and Jay E. Daily, eds., *Encyclopedia of Library and Information Science*, Vol. 21, p. 448. New York: Marcel Dekker, 1977.

Edwards, Susan. "Patents: An Introduction." *Special Libraries* 69 (February 1978): 45-50.

Finley, Ian F. *Guide to Foreign-language Printed Patents and Applications.* London: Aslib, 1969.

Harris, Stafford. "Patent and Trademark Literature." In Frances E. Kaiser, ed., *Handling of Special Materials in Libraries*, pp. 26-45. New York: Special Libraries Association, 1974.

Kaback, Stuart M. "Retrieving Patent Information Online." *Online* 2 (January 1978): 16-25.

Kase, Francis J. *Foreign Patents: A Guide to Official Patent Literature.* Dobbs Ferry, NY: Oceana Publications, Inc., 1972.

McDonnell, Patricia M. "ICIREPAT and International Developments in Patent Information Retrieval." *Special Libraries* 66 (March 1975): 133-9.

Maynard, John T. "How to Read a Patent." *IEEE Transactions on Professional Communication* PC-22 (June 1979): 112-18.

Morehead, Joe. "Of Mousetraps and Men: Patent Searching in Libraries." *The Serials Librarian* 2 (Fall 1977): 5-11.

Newby, Frank. *How to Find out about Patents.* New York: Pergamon Press, 1967.

Ollerenshaw, Kay. "How to Perform a Patent Search: A Step by Step Guide for the Inventor." *Law Library Journal* 73 (Winter 1980): 1-16.

Patents and Trademarks (Subject Bibliography SB-021). Washington, DC: U.S. Government Printing Office, 1982.

Quarda, Gerhard. "The Computerized Patent Documentation System of INPADOC." *International Forum on Information Documentation* 4 (October 1979): 26-35.

U.S. Department of Commerce. Patent and Trademark Office. *General Information Concerning Patents.* Washington, DC: U.S. Government Printing Office, 1979 (C21.2:P27/979).

U.S. Department of Commerce. *Q & A about Patents.* Washington, DC: U.S. Government Printing Office, 1978.

U.S. Small Business Administration. *Introduction to Patents* (Management Aids 240). Washington, DC: U.S. Government Printing Office, 1979.

Woodburn, Henry M. *Using the Chemical Literature: A Practical Guide.* New York: Marcel Dekker, 1974.

7

SCIENTIFIC TRANSLATIONS

INTRODUCTION

After the Second World War, it became increasingly evident that scientists not only had to cope with the exponentially growing scientific literature but also with the fact that a significant portion of the literature is published in languages they cannot read. The UNESCO report on the language problem lamented that " ... leaving qualitative differences aside, the greater part of what is published is inaccessible to most of those who could otherwise benefit from it" since "at least 50 percent of scientific literature is in languages which more than half the world's scientists cannot read."[1]

For English-speaking scientists, the emergence of the Russian language as an important vehicle of scientific information complicates the foreign language problem. Tybulewicz estimated that Russian language output in physics, chemistry, biology, and mathematics is second to that of English with 17, 23, 10, and 21 percents respectively.[2] The number of Russian papers in these subject areas is greater than the combined total of French and German papers.

Table 7.1 (page 112) shows that while the percentage of German and French journal articles abstracted in *Chemical Abstracts* is decreasing, the percentage of Russian publications is relatively stable with 20 percent of total publications.[3] Another evidence of the foreign language problem for English-speaking scientists is that over 50 percent of all science and technology periodicals are published in non-English languages.[4] Similarly, a significant number of books in pure and applied sciences are published in countries such as the USSR and Japan.[5]

TABLE 7.1
LANGUAGE OF PUBLICATION OF JOURNAL LITERATURE ABSTRACTED IN *CHEMICAL ABSTRACTS*
(Percentages by Year)

Language	1961	1966	1972	1978
English	43.3	54.9	58.0	62.8
Russian	18.4	21.0	22.4	20.4
German	12.3	7.1	5.5	5.0
Japanese	6.3	3.1	3.9	4.7
French	5.2	5.2	3.9	2.4
Polish	1.9	1.8	1.2	1.1
Italian	2.4	2.1	0.8	0.6
Czech	1.9	0.9	0.6	0.5
Others	8.3	3.9	3.7	2.5

The significance of the foreign language scientific information was finally recognized in the United States when the Russians surprised the world by launching the Sputnik satellite in 1957. That event forced the realization that scientific and technical information activities should include mechanisms to deal with the information published in foreign languages. Translation of scientific and technical articles into English, of course, is one of the mechanisms available to cope with the foreign language problem.

The U.S. government has actively encouraged the scientific translation activities through its support of 1) cover-to-cover translation of scientific journals, 2) ad hoc and selective translation programs, 3) the production of English language abstracts of foreigh language materials, and 4) dissemination of information about the availability of scientific translations.

COVER-TO-COVER TRANSLATIONS

Cover-to-cover translation involves the complete translation of a journal which is selected on the basis of its importance and/or marketability.[6] The advantage of a cover-to-cover translation is that "it insures against inadvertent omission of significant information, and it facilitates bibliographic handling and searching since issues of the translated periodical can be correlated exactly with issues of the original publication."[7]

In the heyday of federal support for cover-to-cover translations, NSF alone supported a great number of cover-to-cover translations. In 1961, for example, the NSF Foreign Science Information Program supported "the translation of 40 Russian scientific journals on a cover-to-cover basis and eight on a selective translation basis, through some 20 societies and institutions."[8] Some of these translation journals continue to be published even after the federal support ceased. The federal agencies employed two methods for supporting cover-to-cover translations:

1. Providing grants to nonprofit scientific societies or research organizations, and
2. Directly contracting with translating agencies.[9]

The National Science Foundation followed the first method, supporting organizations such as the American Institute of Biological Sciences and the American Institute of Physics. Examples of translated journals issued under the sponsorship of the National Science Foundation include

Chinese Physics,
Doklady Akademii Nauk SSSR: Otdel Biokhimii,
Pochvovedenie,
Astronomicheskii Zhurnal,
Geokhimiia, and
Radiotekhnika.

The National Institutes of Health (NIH), in contrast to the National Science Foundation, preferred contracting with translating agencies such as the Consultants Bureau and Pergamon Institute. In this case, NIH assumed the responsibility of selecting the periodicals to be translated and then exercising the quality control. Examples of the NIH sponsored periodicals include

Biokhimiia,
Biofizika,
Voprosy Onkologii, and
Voprosy Virusologii.

National Technical Information Service is the source of many translated journals. Examples of such journals are

Ukrainian Physics Journal (Cover-to-cover translation of *Ukrainskii Fizicheskii Zhurnal.* Sponsored in part by National Science Foundation Special Foreign Currency Science Information Program. Issued by Department of Energy.)

Air Conservation (Environmental Protection Agency. Translation of *Ochrona Powietrza* (Poland). Sponsored in part by NSF Special Foreign Currency Science Information Program.)

Radiobiology (E1.44:(v.nos.&nos.)) (Joint Publications Research Service. Translation of *Radiobiologiia* (USSR). Sponsored in part by Department of Energy, Oak Ridge, Tennessee Technical Information Center)

Space Biology and Aerospace Medicine (PrEx7.22/12:(nos.)) (Joint Publications Research Service. Translation of *Kosmicheskaya Biologiya i Aviokosmicheskaya Meditsina.*)

AD HOC AND SELECTIVE TRANSLATIONS

Ad hoc translations, unlike cover-to-cover translations of journals, are translations of individual articles or chapters of books, usually selected by individual scientists or organizations.[10] Selective translations, on the other hand, "are collections of translated articles selected by editor or publishers from one or more journals, usually on the same or similar subjects, and published in a selective translation journal."[11] Joint Publications Research Service (JPRS) and the National Science Foundation Special Foreign Currency Science Information Program, among others, are active in producing these types of translations.

Of the more than eighty-five thousand reports translated by JPRS (1000 North Glebe Road, Arlington, VA 22201) since 1957, about half are in science and technology.[12] Although originally established to serve the translation needs of the Central Intelligence Agency, JPRS conducts translation work for many federal agencies that need "English-language translations of specific books, newspapers and periodical articles, and similar materials."[13]

JPRS issues a number of selections of translations dealing with broad subject areas. Examples of such selections are

Worldwide Report:
- a) Environmental Quality (PrEx7.14:(nos.))
- b) Epidemiology (PrEx7.14/2:(nos.))
- c) Nuclear Development and Proliferation (PrEx7.14/4:(nos.))
- d) Telecommunications Policy, Research, and Development (PrEx7.14/5:(nos.))

China Report:
- a) Agriculture (PrEx7.15:(nos.))
- b) Science and Technology (PrEx7.15/6:(nos.))

For example, *Translations on Eastern Europe: Scientific Affairs*, a predecessor to the current series called *East Europe Report: Scientific Affairs* (PrEx7.17/2: [nos.]), "contains press and radio coverage on the development of and progress in the various theoretical and applied scientific disciplines and technical fields; and the administration, structure, personnel, and research plans of leading East European scientific organizations and institutions, particularly the academies of sciences."[14]

One of the reports in this series (JPRS Report No. 73816, July 9, 1979) contained seven articles on East Europe, Bulgaria, Hungary, and Romania. These articles dealt with development of computer systems in East Europe, Bulgarian activities in space research, electronics and computer industries; Romanian plans for solar energy installations, and an interview with A. M. Prokhorov, a Nobel Laureate, on Hungarian-Soviet scientific cooperation.

The Special Foreign Currency Science Information Program (SFCSI) was established pursuant to a law authorizing the President of the United States to use the surplus foreign currencies to

> collect, collate, translate, abstract, and disseminate scientific and technological information and conduct research and support scientific activities overseas including programs and projects of scientific cooperation between the United States and other countries....
>
> (72 Stat 275)

The responsibility for administering and coordinating translation activities has been assigned to the National Science Foundation. The SFCSI Program is "primarily designed to supplement the information needs of U.S. Government research scientists."[15]

The translations of SFCSI Program are selected by federal scientists "acting on recommendations from scientists, academic institutions, and professional societies."[16] The program produces more than fifty thousand pages per year and, since 1959 has translated over one million pages of original language material and made it available to English-speaking scientists.[17]

There are numerous sources that provide access to the ad hoc and selected translations sponsored by the U.S. government. JPRS translations are announced in the *Government Reports Announcements & Index* and *Monthly Catalog of U.S. Government Publications*. Indexes to JPRS reports (by keyword, author, personal name, title, and series) are available from Bell & Howell in the form of a serial publication: *Transdex*. JPRS reports are distributed to GPO depository libraries. In addition, JPRS reports can be ordered from the National Technical Information Service. Finally, access to the JPRS reports is also provided by the Readex Microprint Corporation which makes them available in microcard format.

SFCSI translations are listed in *SFCSI List of Translations in Process*, an annual publication available free from National Technical Information Service. Only a part of SFCSI translations are available from NTIS. Others are available from the sponsoring agencies or from the National Translations Center. Many federal ad hoc translations are listed in the National Translations Center's *Translations Register-Index*.

ENGLISH-LANGUAGE ABSTRACTS OF FOREIGN LANGUAGE MATERIAL

The third approach followed in dealing with the foreign language problem is 1) the coverage of foreign language literature through normal indexing and abstracting services and 2) the JPRS abstracting serials which provide English-language abstracts of foreign language materials.

An example of early attempts at improving the coverage of foreign language material by indexing and abstracting services was the support provided by the federal government to BioSciences Information Services and Excerpta Medica Foundation. Excerpta Medica Foundation published *Abstracts of Japanese Medicine* (1960-1962). The Special Foreign Currency Science Information (SFCSI) program cooperated with *Biological Abstracts* in covering Russian biological sciences literature. The National Library of Medicine supported a publication called *Drug Digests from the Foreign Language Literature* which was published from 1964 to 1968.

Currently, many federally published indexing and abstracting services cover foreign language literature in their respective subject areas. For example, about 20 percent of the abstracts included in *Carcinogenesis Abstracts*,* a publication sponsored by the National Cancer Institute, are of foreign language items.

*Ceased publication.

Abstractors for this publication are fluent in languages such as Russian, German, and Japanese.

Joint Publications Research Service has undertaken the task of providing English-language abstracts of foreign language scientific literature. For example, the JPRS series titled *USSR Report: Cybernetics, Computers and Automation Technology* contains "articles, abstracts, and news items on theory, design, development and application of analog and digital apparatus, elements and components of control systems, reliability and optimality, information theory, and the theory of automata."[18] Other abstract series issued by JPRS are:

USSR Report:
- a) Chemistry (PrEx7.22/4:(nos.))
- b) Cybernetics, Computers, and Automation Technology (PrEx7.22/5:(nos.))
- c) Electronics and Electrical Engineering (PrEx7.22/6:(nos.))
- d) Engineering and Equipment (PrEx7.22/7:(nos.))
- e) Earth Sciences (PrEx7.22/8:(nos.))
- f) Space (PrEx7.22/9:(nos.))
- g) Materials Science and metallurgy (PrEx7.22/10:(nos.))
- h) Physics and Mathematics (PrEx7.22/11:(nos.))

DISSEMINATION OF INFORMATION ABOUT TRANSLATIONS

One approach followed by the federal government in disseminating information about the availability of translations was the funding of nonprofit organizations such as the National Translations Center in support of their attempts to acquire, index, and distribute translations irrespective of their origin. The National Translations Center was originally known as SLA Translations Center. Established in 1953 by the Special Libraries Association and operated by John Crerar Library, the center was organized to eliminate "costly duplication of translating effort" and "to disseminate – quickly and nominally – copies of the translations in its collection."[19]

In this effort, the center received support from the National Science Foundation, U.S. Public Health Service, and Clearinghouse for Federal Scientific and Technical Information. In 1968 the center became National Translations Center under the sponsorship of John Crerar Library and was once again supported by the National Science Foundation. Similarly, the center was funded by the foundation to compile the *Consolidated Index of Translations in English* (New York: Special Libraries Association, 1969) which included information on about one hundred forty-two thousand translations. The federal government's support to the center continued into 1973.

The federal government also disseminates information about translations by announcing them in various bibliographies. Examples of such publications are *Technical Translations* (1959-1966), *Bibliography of Medical Translations, Bibliography of Translations from Russian Scientific and Technical Literature* (1953-1966), and *Recent Translations: A Selected List*. A more recent example is: *Translated Tables of Contents of Current Foreign Fisheries, Oceanographic, and Atmospheric Publications* (C55.325:(nos.))

Similarly, *Monthly Catalog* and *Government Reports Announcements & Index* serve as current, albeit incomplete, sources of information on the availability of translations from the federal government. Ildiko D. Novak, Chief of National Translations Center, writes:

> The inclusion of translations in *Government Reports Announcements and Index* is so minimal that it becomes almost non-existent. We noticed that translations of journal articles are completely excluded, even though these comprise the bulk of all government-produced translations.[20]

Although many federally produced translations are neither announced nor made available to the public, Mrs. Novak notes that a number of federal agencies such as Environmental Protection Agency, National Institutes of Health, Departments of Energy and Agriculture do deposit their translations with the National Translations Center.[21] In fact, Mrs. Novak believes that, as time passes, "The Center is becoming more and more the sole depository for many federally produced translations."[22]

REFERENCES

1. United Nations Educational, Scientific and Cultural Organization, *Scientific and Technical Translating and Other Aspects of the Language Problem* (Paris: 1957), p. 13.
2. A. Tybulewicz, "Cover-to-Cover Translations of Soviet Scientific Journals," *Aslib Proceedings* 22 (1970): 55-62.
3. American Chemical Society, Chemical Abstracts Service, *CAS Today: Facts and Figures about Chemical Abstracts Service* (Columbus, OH, 1980), p. 30.
4. D. N. Wood, "The Foreign-Language Problem Facing Scientists and Technologists in the United Kingdom—Report of a Recent Survey," *Journal of Documentation* 23 (June 1967): 117-30.
5. United Nations, Department of International Economic and Social Affairs, Statistical Office, *Statistical Yearbook: 1977* (New York, 1978), Tables 213, 214.
6. James D. Anderson, "Ad Hoc and Selective Translations of Scientific and Technical Journal Articles: Their Characteristics and Possible Predictability," *Journal of the American Society for Information Science* 29 (May 1978): 130-35.
7. U.S. National Science Foundation, Office of Scientific Information, *Providing U.S. Scientists with Soviet Scientific Information* (Washington, DC, 1958), p. 3.
8. Paul S. Feinstein, "Translation Activities of the Foreign Science Information Program," *Special Libraries* 53 (January 1962): 26-29.
9. *Providing U.S. Scientists with Soviet Scientific Information*, op. cit., p. 4.

10. Anderson, op. cit., p. 130.

11. Ibid.

12. U.S. National Technical Information Service, *NTIS Information Services* (General Catalog No. 6 for North America) (Springfield, VA, 1979), p. 22.

13. Rita Lucas and George Caldwell, "Joint Publications Research Service Translations," *College and Research Libraries* 25 (March 1964): 103-10. See also David Y. Allen, "Buried Treasure: The Translations of the Joint Publications Research Service," *Government Publications Review* 9 (1982): 91-98.

14. U.S. Joint Publications Research Service, *Translations on Eastern Europe: Scientific Affairs* (JPRS Report Number 73816) (Arlington, VA, 1979).

15. National Science Foundation, Special Foreign Currency Science Information Program, *List of Translations in Process: 1979* (Springfield, VA: National Technical Information Service, 1979), p. i. See also Charles Zalar, "United States Special Foreign Currency Science Information Program," in *Expanded Communications in a Shrinking World* (Washington, DC: Special Libraries Association, 1968), pp. 49-54.

16. *NTIS Information Services*, op. cit., p. 16.

17. *List of Translations in Process: 1979*, op. cit., p. i.

18. U.S. Joint Publications Research Service. *USSR Report: Cybernetics, Computers, and Automation Technology* (JPRS Report Number 82695) (Arlington, VA, 1983).

19. Mary L. Allison, "The SLA Translations Center," in *Bowker Annual of Library and Book Trade Information: 1966* (New York: Bowker, 1966), pp. 121-23.

20. Personal communication from Mrs. Ildiko D. Novak, Chief, National Translations Center (June 23, 1980).

21. Ibid.

22. Ibid.

8

STANDARDS AND SPECIFICATIONS*

INTRODUCTION

Standardization, according to the International Organization for Standardization (ISO), is "the process of formulating and applying rules for an orderly approach to a specific activity for the benefit and with the cooperation of all concerned and in particular for the promotion of optimum overall economy taking due account of functional conditions and safety requirement."[1] Standardization is fundamental to many aspects of modern life including science, technology, industry, commerce, health, and education.

Standards and specifications are documents that stipulate or recommend: 1) minimum levels of performance and quality of goods and services, and 2) optimal conditions and procedures for operations in science, industry, and commerce, including production, evaluation, distribution, and utilization of materials, products, and services. Standards are established by general agreement among representatives of consumers, designers, manufacturers, distributors, and other concerned groups.

Standards and specifications are important components of technical literature. Thousands of standards and specifications are used by scientists, engineers, technologists, designers, manufacturers, and consumers of virtually

*The author of this chapter is K. Subramanyam, Associate Professor, School of Library and Information Science, Drexel University, Philadelphia, Pennsylvania.

every type of material, product, or service ranging from common items of everyday use such as food, drugs, and toys, to extremely complicated equipment and components used in space vehicles and nuclear reactors.

Standards are essential in scientific research to ensure reproducibility of research and accuracy and reliability of the results of research. In industrial and commercial practice standards are essential in order to 1) prevent avoidable wastage of resources and manpower, 2) to enhance safety, speed, and productivity, 3) to ensure uniformity, reliability, and excellence of product quality, and 4) to achieve overall efficiency and economy.

A product standard provides a basis for evaluating a product's acceptability based on its performance under stated conditions or tests. Typically, a product standard addresses several attributes of a product. A standard relating to building materials might, for example, specify the engineering properties, weather resistance, and fire resistance of the material. Some product standards provide a "pass" or "fail" rating; other standards contain a multipoint or continuous grading system that facilitates measurement and comparison of degrees of performance of products.

Manufacturers use product standards and specifications so that their processes and products 1) can be certified by certifying agencies (such as the National Bureau of Standards), 2) can be accepted by government procurement agencies such as the Federal Supply Service of the General Services Administration, or 3) conform to the regulations of regulatory agencies such as the Environmental Protection Agency or the Consumer Product Safety Commission. Product standards are also useful to buyers and enable them to buy products with the confidence that the products meet minimum performance or quality levels prescribed by standards.

Standardization is a continuous process. Standards are amended or revised frequently to keep up with changes in technology or accepted practices and needs, and to reaffirm the continued relevance and applicability of the standards for the purpose for which they were developed. The American National Standards Institute (ANSI) reviews its standards once in five years; such reviews may result in 1) reaffirmation of the standards, or 2) minor or substantive changes in the standards to meet the changing needs.

DEFINITIONS

The term "standards" encompasses a wide range of technical documents variously referred to as specifications, tests and test methods, analyses, assays, reference samples, recommended practices, guides to good practice, nomenclature, symbols, grading rules, codes, forms and contracts, criteria, methods and codes of practice.

Although the terms "standard," "specification," and "standard specification" are often used interchangeably, there are differences between a standard and a specification. The ISO defines a standard as the result of a particular standardization effort, approved by a recognized authority; it may take the form of 1) a document containing a set of conditions to be fulfilled or 2) a fundamental unit or physical constant (e.g., ampere, meter, or kilogram).

A standard is defined by ANSI as "A standard is a specification accepted by recognized authority as the most practical and appropriate current solution to a recurring problem." A specification, according to ANSI "is a concise statement of the requirement for a material, process, method, procedure, or service including, whenever possible, the exact procedure by which it can be determined that the conditions are met within the tolerances specified in the statement; a specification does not have to cover specifically recurring subjects or subjects of wide use, or even existing objects."

Although both standards and specifications stipulate acceptable levels of dimensions, quality, performance, or other attributes of materials, products, and processes, the scope of applicability of a standard is usually much greater than that of a specification. For example, a state government agency awarding a contract for the construction of a school building may specify to its contractor the minimum level of illumination required in the various parts of the proposed building. Such a statement is incorporated in a "specification."

On the other hand, the recommendations for minimum levels of brightness in school buildings all over the state, or the entire country, may be developed by representatives of academic institutions, government agencies, technical societies, and industrial organizations, acting in concert. Such a recommendation, when approved by an appropriate standards body (e.g., ANSI), becomes a "standard."

A specification may be thought of as a purchase document that contains the description of the technical features of a product, material, process, or service that are required to meet the specific needs of a purchaser or an industry.[2] A standard consists of similar descriptions of technical features, formulated by agreement, authority, or custom, but applicable to a broader range of situations, at a corporate, national, or international level.

TYPES OF STANDARDS AND SPECIFICATIONS

As mentioned before, the term "standards and specifications" covers a variety of documents including standards, specifications, codes of practice, recommendations, guidelines, terminology and symbols, and so on. Though several distinct types of standards and specifications may be recognized on the basis of their purpose or formulating agency, many standards and specifications are composite in nature and possess the characteristics of more than one type. Various kinds of specifications, including dimensional specifications, "technique" specifications, product specifications, performance specifications, packaging specifications, etc., have been described by Reeves.[3] Based on their purpose, standards may be categorized into the following major types:[4, 5]

1. *Dimensional Standards.* These specify standard dimensions to achieve interchangeability in assembly of components in manufacturing industries and other applications, and to facilitate replacement of worn or damaged parts; an example is ISO 1020-1969: "Dimensions of Daylight Loading Spools for Double 8mm Motion Picture Film."

2. *Material Standards.* Materials standards specify the composition, quality, chemical or mechanical properties of materials such as alloys, fuels, paints, etc.; for example, "Age control of age-sensitive elastomeric material," MIL-STD-1523. Another example of a voluntary product standard is ANSI/ASTM D 1655-77: "Standard Specification for Aviation Turbine Fuels."

3. *Performance Standards.* A performance standard specifies the minimum performance or quality of a product or component so that it can fulfill its intended functions at acceptable levels of efficiency and safety. An example of performance standard is the military standard: MIL-STD-1404A: "Scraper, Earthmoving, Towed, Hydraulic-operated."

4. *Standards of Test Methods.* These recommend conditions, procedures, and tools for testing, evaluating and comparing the quality or performance of materials and products; an example is ANSI B38.1-1970: "Methods of Testing for Household Refrigerators, Combination Refrigerator-Freezers, and Household Freezers."

5. *Codes of Practice.* These specify procedures for installation, operation, maintenance, and other industrial operations to achieve safety and uniformity in such operations; an example is CP 1021:1973: "Code of Practice for Cathodic Protection."

6. *Standards of Terminology and Graphic Symbols.* Terminology and symbols used in technical communication, drawings, and flowcharts are also standardized to enable engineers and technologists to communicate unambiguously; examples are IEEE Standard 100-1972: "IEEE Standard Dictionary of Electrical and Electronics Terms" (716p.), and ANSI Y14.2-1973: "Engineering Drawing and Related Documentation Practices: Line Conventions and Lettering." Standard nomenclature for chemical compounds recommended by the International Union of Pure and Applied Chemistry (IUPAC) and similar agencies may also be considered in this category.

7. *Documentation Standards.* These are standards for the layout, production, reproduction, distribution, classification, indexing, and bibliographic description of documents. Committee Z39 of ANSI and Technical Committee 46 of ISO are concerned with developing such standards. Examples are ANSI/ASTM E250-76: "Standard Recommended Practice for Use of CODEN" and ANSI Z39.18-1974: "American National Standard Guidelines for Format and Production of Scientific and Technical Reports."

8. *Standard Fundamental Units.* Fundamental standards deal with the measurement of length, mass, time, temperature, various forms of energy, force, and other quantifiable fundamental entities that are basic to all scientific and technical endeavors. Units such as calorie, ampere, gram, meter, second, etc. are all precisely defined by standards so that measurements in science and industry can be performed with a high degree of accuracy.

9. *Standards for Uniformity and Quality.* Hemenway has identified two principal types of product standards, based on their function: 1) standards for uniformity and 2) standards of quality.[6] Standards for uniformity are created to achieve simplicity by reducing needless variety in product sizes, speeds, and other attributes of products.

The second function of standards for uniformity is to facilitate interchangeability of products and components manufactured by different companies. Standards for uniformity are helpful in specifying dimensions and other features of products, e.g., electric lamp sockets, screw threads, etc.

Standards for railroad track gauge also enforce uniformity. There are various track gauges in use in different countries: 5 ft. 6 in. in South America, 4 ft. 8½ in. in the United States, and 3 ft. 6 in. in South Africa. The tracks on the original Great Western Railroad were 7 feet apart. In any given railroad system, uniformity of track gauge is extremely important to ensure safety of passengers and interchangeability of tracks, locomotive engines, and rolling stock.

Quality standards usually specify the minimum levels of product quality or performance, and are helpful in assessing the quality of products. The United States Department of Agriculture requires that commercial peanut butter must contain at least 90 percent peanuts by weight, or that orange drink must contain at least 10 percent orange juice; these are examples of product quality standards.

10. *Voluntary Standards.* Standards developed by private (i.e., non-government) agencies are sometimes called voluntary standards, as opposed to the mandatory standards or regulations issued by government regulatory agencies such as the Environmental Protection Agency or the Nuclear Regulatory Commission. But many of these voluntary standards are adopted by government agencies for procurement or regulatory purposes, and thus eventually become mandatory standards. Municipal and state regulatory agencies often require suppliers of products and services to conform to private standards.

Federal government agencies also rely extensively on private standards. Federal regulatory agencies such as the Consumer Product Safety Commission, the Occupational Safety and Health Administration and the Department of Housing and Urban Development, the Enrivonmental Protection Agency, and the Food and Drug Administration rely fairly extensively on private standards.

Federal government agencies such as the Department of Defense also use private standards. Guidelines developed by the Interagency Committee on Standards Policy and the National Commission on Government Procurement are expected to increase reliance on private standards by federal government agencies for both regulatory and procurement purposes.[7]

SOURCES OF STANDARDS AND SPECIFICATIONS

Standards and specifications are formulated by 1) companies, 2) trade associations and technical societies, 3) government agencies, 4) national standardization organizations, and 5) international standardization organizations. Government agencies are among the largest producers and users of standards and specifications in most countries. The United States government is the world's largest bulk purchaser of virtually every kind of commodity and service, and predictably, it is also one of the largest producers and users of standards and specifications.

This chapter concentrates on sources of standards and specifications within the federal government, and only briefly deals with the standardization activities of ANSI; other sources of standards and specifications are described elsewhere.[8, 9]

Most industrially advanced countries and many developing countries have a national standards organization to prepare, approve, and publish standards and to coordinate standardization effort at a national level. In the United States, ANSI is the national agency responsible for promoting and coordinating the standardization effort in the private sector. Although ANSI is not a government agency, a brief discussion of its activities will not be out of place here in view of the facts that a great number of voluntary standards made or approved by ANSI are adopted and used by various government agencies and ANSI is the official representative of the United States viewpoint in international standardization forums.

THE AMERICAN NATIONAL STANDARDS INSTITUTE (ANSI)

ANSI's predecessor, the American Engineering Standards Committee, was established in 1918 by the American Society for Testing and Materials (ASTM) and four other engineering societies: The American Society of Mining and Metallurgical Engineers, the American Institute of Electrical Engineers, the American Society of Mechanical Engineers, and the American Society of Civil Engineers. The U.S. Departments of Commerce, War, and Navy were cofounders. In 1928 the name of the committee was changed to American Standards Association. From 1966 until 1969 the organization was known by the name United States of America Standards Institute. The present name was adopted in October 1969.

ANSI is a nonprofit federation with a membership of more than 160 technical and trade associations and professional societies, about 1,000 companies, and a few consumer associations.

The major functions of ANSI are

to coordinate national standardization activity by reducing duplication, overlapping, and variations in standards.

to facilitate the development of new standards and the revision or amendment of existing standards.

to approve standards developed by other agencies (e.g., professional societies, industries, etc.) as American National Standards.

to represent United States interests in international standards bodies. ANSI is the U.S. member of the International Organization for Standardization (ISO), the International Electrotechnical Commission (IEC), the Pan-American Standards Commission, and other international standards bodies.

to serve as a clearinghouse for information on American, foreign, and international standards, and to sell copies of these standards.

The following publications of ANSI are important:

ANSI Catalog, published annually, contains a complete listing of all currently valid American National Standards, as well as standards of ISO and IEC.

Standards Action, a biweekly newsletter, summarizing the standardization activities of ANSI including standards submitted to ANSI for approval, standards approved or withdrawn, newly published standards, and reports of American National Standards committees.

ANSI Reporter, a general, biweekly newsletter.

Copies of ANSI standards and also international standards can be purchased from the American National Standards Institute, 1430 Broadway, New York, NY 10018.

THE NATIONAL BUREAU OF STANDARDS (NBS)

A key government agency dedicated to standardization and research at a national level is the National Bureau of Standards. NBS was established by an act of Congress on March 3, 1901 with the broad objective of strengthening and advancing the nation's science and technology and facilitating their application for public benefit. Rexmond Cochrane has written a comprehensive history of the NBS including an account of the developments that led to its inception at the turn of the century.[10] The primary mission of the bureau is to provide national leadership in the development and use of accurate, uniform techniques for measurement which are very crucial for progress in science, technology, and industry.

The bureau provides

a foundation for the nation's physical measurement system.

scientific and technological services to industry and government.

a technical basis for equity in trade.

technical services to promote public safety.

In addition, the bureau is also responsbile for

the development and sale of standard reference materials required for the calibration of measuring instruments.

assisting industries in developing and publishing voluntary product standards under procedures published by the Department of Commerce in Part 10, Title 15 of the *Code of Federal Regulations* (GS4.108:Title 15/Parts 300-399).

The bureau's technical work is performed by three units: The National Measurement Laboratory, the National Engineering Laboratory, and the Institute for Computer Sciences and Technology. Each of these units consists of several research centers.

The responsibilities of the National Measurement Laboratory include

provision of a national system of physical, chemical, and materials measurement.

coordination of the national measurement system with similar systems of other countries.

provision of technical services to the nation's scientific, industrial, and commercial communities to promote accurate and uniform physical and chemical measurement.

conducting research leading to improved methods of measurement, standards, and data on properties of materials.

provision of advisory and research services to other government agencies.

development, production, and distribution of standard reference materials (see also NSRDS in this chapter).

provision of calibration services.

The National Measurement Laboratory consists of the following research centers: Absolute Physical Quantities, Radiation Research, Thermodynamics and Molecular Science, Analytical Chemistry, and Materials Science.

The National Engineering Laboratory has the following responsibilities:

provision of technology and technical assistance to both the public and private sectors.

conducting research in engineering and applied science to support the above service.

development of engineering data and measurement capabilities.

development of test methods.

proposing new engineering standards and codes and modifications to such standards and codes.

The National Engineering Laboratory consists of the following research centers: Applied Mathematics, Electronics and Electrical Engineering, Mechanical Engineering and Process Technology, Building Technology, Fire Research, Consumer Product Technology, and Field Methods.

The objectives of the Institute for Computer Sciences and Technology are to conduct research and provide scientific and technical services to federal agencies in the selection, acquisition, and application of computer technology to improve government operations according to Public Law 89-306 (40 U.S.C. 759). The institute performs the following functions:

managing the Federal Information Processing Standards program.

developing federal ADP standards and guidelines.

managing federal participation in ADP voluntary standardization activities.

providing the technical foundation for computer-related policies of the U.S. government.

The Institute for Computer Sciences and Technology consists of two research centers: Programming Science and Technology, and Computer Systems Engineering.

The publication program of NBS is quite extensive and encompasses a wide range of standards and research literature. In 1976, its seventy-fifth anniversary year, NBS published a total of nearly one hundred ten thousand pages. About 79 percent of this material was published in the bureau's own publications; the other 21 percent appeared in non-NBS journals, books, and proceedings.

All the publications of NBS, including those published in non-NBS journals, books, and proceedings, are listed in an annual catalog, e.g.

Publications of the National Bureau of Standards. 1981 Catalog (National Bureau of Standards Special Publication 305, Supplement 13). Washington, DC: National Bureau of Standards, 1981 (C13.10:305/13).

Most of the monographs, reports, and technical notes are announced also in the *Monthly Catalog of United States Government Publications* and are distributed by the Superintendent of Documents, Government Printing Office, Washington, DC 20402.

Following are the principal eleven publications of NBS:

1. *The Journal of Research of the National Bureau of Standards* (C13.22:(v.nos.&nos.)), issued six times a year, reports NBS research and development in physics, chemistry, engineering, mathematics and computer science, with emphasis on measurement methodology and technology underlying standardization. Each issue contains a list of all recent NBS publications, including those published in non-NBS media.

2. *Voluntary Product Standards* developed under procedures established by the U.S. Department of Commerce (published in Part 10, Title 15 of the *Code of Federal Regulations*). The voluntary product standards program is administered by the NBS as a supplement to the product standardization activities in the private sector. The following are two examples of voluntary product standards:

> Carbonates Soft Drink Bottles (ANSI/VPS PS 73-77). National Bureau of Standards Product Standard 73-77. 11 p. November 1977 (C13.20/2:73-77).
>
> Toy Safety (ANSI/VPS PS 72-76). National Bureau of Standards Product Standard 72-76. 30p. January 1977 (C13.20/2:72-76).

3. *Monographs* are major contributions to the technical literature on various subjects related to the bureau's research and development activities. The following examples are typical of NBS monographs:

> B. W. Steiner. *An Institutional Plan for Developing National Standards with Special Reference to Environment, Safety, and Health.* NBS Monograph No. 165. September 1979. 24p. (C13.44:154).
>
> J. P. Montgomery and D. C. Chang. *Electromagnetic Boundary-Value Problems Based upon a Modification of Residue Calculus and Function Theoretic Techniques.* NBS Monograph No. 164. June 1979. 183p. (C1344:164).

4. *Technical Notes* are reports on technical topics, "analogous to monographs, but not so comprehensive in scope or definitive in treatment of the subject area." Examples of technical notes are

> S. J. Tredo et al. *An Investigation of Air Infiltration Characteristics and Mechanisms for a Townhouse.* NBS Technical Note TN 992. 36p. August 1979 (C13.46:992).
>
> C. A. Hoer. *Calibrating a Six-Port Reflectomer with Four Impedance Standards.* NBS Technical Note TN 1012. 24p. March 1979 (C13.46:1012).

5. *Handbooks* contain recommended codes of engineering and industrial practice, including safety codes, developed in cooperation with interested industries, professional organizations, and regulatory bodies. Examples of handbooks are

ANS N538. *Classification of Industrial Ionizing Radiation Gauging Devices* (ANSI N538-1979). NBS Handbook No. 129. October 1979. 29p. (C13.11:129).

Specifications, Tolerances, and Other Technical Requirements for Weighing and Measuring Devices. 1979 edition. NBS Handbook No. 44. December 1979. 213p. (C13.11:44/6).

6. *Special Publications* include NBS annual reports, bibliographies, and proceedings of conferences sponsored by NBS. The following publications are typical of NBS Special Publications:

Alexander J. Glass and Arthur H. Guenther. *Laser Induced Damage in Optical Materials: 1978.* Proceedings of a symposium sponsored by National Bureau of Standards, September 12-14, 1978, Boulder, Colorado. NBS Special Publication 541. December 1978 (C13.10:541).

S. A. Berry, ed. *Proceedings of the National Conference on Regulatory Aspects of Building Rehabilitation.* Proceedings of a conference held at the National Bureau of Standards, Gaithersburg, Maryland, October 30. 1978. NBS Special Publication SP549. August 1979. 220p. (contains 18 papers presented at the conference) (C13.10:549).

7. *Applied Mathematics Series* includes studies in applied mathematics as well as mathematical tables and manuals.

8. *Building Science Series* presents research results, test methods, and performance criteria of structures, building elements, systems and materials.

9. *Consumer Information Series* contains practical information of interest to the consumer presented, with illustrations, in an easily understandable language.

All of the above publications are obtainable from the Superintendent of Documents, Government Printing Office, Washington, DC 20402.

The following publications are available from the National Technical Information Service, Springfield, VA 22161:

10. *Fed al Information Processing Standards Publications* (C13.52:(nos.)) (FIPS PUB), collectively constituting the Federal Information Processing Standards Register. The register contains information on standards issued by the NBS pursuant to the Federal Property and Administrative Services Act of 1949, as amended, Public Law 89-306 (79 Stat. 1127).

11. *NBS Interagency Reports* (*NBSIR*) is a special series of interim reports on work performed at NBS for outside sponsors (both government and non-government). Copies of these also may be obtained from the sponsoring agencies concerned.

The following indexes prepared by the Standards Information Service of NBS are also available from NTIS:

Index of International Standards. NBS Special Publication 390 (1974) (C13.10:390).

An Index of U.S. Voluntary Engineering Standards. NBS Special Publication 329 (1971). Supplements to this index are available from the Superintendent of Documents, U.S. Government Printing Office, Washington, DC 20402 (C13.10:329).

World Index of Plastics Standards. NBS Special Publication 375 (1973) (C13.10:375).

A *Directory of U.S. Standardization Activities* prepared by the Standards Information Service of NBS is also available from the Superintendent of Documents, U.S. Government Printing Office. Further information on the activities, services, and products of NBS can be obtained from the Technical Information and Publications Division, National Bureau of Standards, Washington, DC 20234.

NATIONAL STANDARD REFERENCE DATA SYSTEM (NSRDS)

The importance of critically evaluated physical and chemical data to the scientific community can hardly be overestimated. Several countries including the United States, the United Kingdom, and the USSR have set up national-level organizations to collect, evaluate, organize, and publish vast amounts of numerical data widely scattered in published and unpublished sources. In the United States, the National Standard Reference Data System (NSRDS) was established in May 1963 by the Federal Council for Science and Technology. The aims of the NSRDS are two-fold:

to provide critically evaluated numerical data in a convenient and accessible form to the scientific and technical community; and

to provide feedback into experimental work to help raise the general standards of measurement.

Further impetus to this program was given by the Standard Reference Data Act which became a law on July 11, 1968 (P.L. 90-396). The act states that "it is the policy of the Congress to make critically evaluated data readily available to scientists, engineers, and the general public.... The Secretary of Commerce is authorized and directed to provide or arrange for the collection, compilation, critical evaluation, publication, and dissemination of standard reference data."[11]

NSRDS is administered by the Office of Standard Reference Data at the NBS, and consists of a network of more than twenty-five data centers located throughout the country in government agencies, academic institutions, and laboratories in the private sector. The Office of Standard Reference Data does not operate or directly supervise the activities of the data centers; it coordinates their technical programs pertaining to the generation, collection, evaluation, and dissemination of standard reference data in the following seven program areas:

1. Nuclear properties.
2. Atomic and molecular properties.
3. Solid-state properties.
4. Thermodynamic and transport properties.
5. Chemical kinetics.
6. Colloid and surface properties.
7. Mechanical properties.

The principal output of NSRDS consists of compilations of evaluated data, and is disseminated through the following channels:

The Journal of Physical and Chemical Reference Data, a quarterly journal containing data compilations and critical data reviews, published jointly by the NBS, the American Institute of Physics, and the American Chemical Society.

NSRDS-NBS series of publications distributed by the Superintendent of Documents, U.S. Government Printing Office, Washington, DC 20402.

Publications of professional societies and commercial publishers.

Responses by NSRDS and its data centers to inquiries for specific data.

NSRDS also makes available selected technical data on magnetic tape as well as computer programs on magnetic tape for handling the technical data.

Information on NSRDS publications may be obtained from the Office of Standard Reference Data, National Bureau of Standards, Washington, DC 20234. Partial lists of NSRDS publications may be seen in the following publications:

CRC Handbook of Chemistry and Physics (CRC Press).

Aluri, R., and P. A. Yannarella. "NBS: A Compilation." *Special Libraries* 65 (February 1974): 77-82.

U.S. Department of Commerce. National Bureau of Standards. *Critical Evaluation of Data in Physical Sciences: A Status Report on the NSRDS.* NBS Technical Note 947 (January 1977) (C13.46:947).

National Standard Reference Data System. *Publication List 1964-1979.* LP 81. Washington, DC: U.S. Department of Commerce, National Bureau of Standards, 1979 (C13.37/4:N21/964-979).

NSRDS publications are also listed in the annual catalogs of NBS publications described in the previous section. Current activities and new products of NSRDS are announced in the bimonthly newsletter *NSRDS Reference Data Report* (C13.48/3:v.nos.&nos.) NSRDS publications can be purchased from the Superintendent of Documents, U.S. Government Printing Office, Washington, DC 20402.

THE U.S. DEPARTMENT OF DEFENSE (DoD)

The Department of Defense and its various agencies together constitute one of the largest producers and users of standards and specifications. The number of standards and specifications in the military field alone is estimated to be well over two hundred thousand; however, not all of these standards and specifications are actually produced by the DoD. The department endorses and adopts for its own use standards approved by ANSI, ASTM, or other organizations.

The Defense Standardization Program (DSP) was established in 1953 to continue the standardization work begun during 1916-1921 by the erstwhile Army-Navy Aeronautical Standards Board. The DSP is managed by the Defense Material Specifications and Standards Board. The activities of DSP encompass the wide range of materials, equipment and parts, engineering practices, and processes used by the defense services. The DSP seeks to promote economy, reliability, and effectiveness of resources by controlling the proliferation of items and practices through the development and use of standards and specifications.

Specifications approved by the U.S. Army, Navy, and Air Force are issued in the well-known MIL series of specifications dating back to 1944. The Defense Standardization Act of 1952 provided for the coordination of standardization effort throughout the United States armed forces. An earlier series of specifications known as the JAN series (Joint Army Navy series), which was in use when some of the air forces of the United States were a part of the U.S. Army (USAAF), is now merged into the MIL series.

The designator for a military specification consists of three parts: 1) the initials MIL (or JAN for older specifications), 2) an initial letter of the title, and 3) a serial number, as illustrated in the following examples:

MIL-P-46210 "Plastic Molding and Extrusion Material, Polysulfone."

MIL-L-21260 "Lubricating Oil."

Military standards are similarly designated, except that the second component consists of the letters STD instead of the first letter of the title:

MIL-STD-698A "Quality Standard for Aircraft Pneumatic Tire and Inner Tubes."

MIL-STD-1256A "Rubber-Coated Parts for Machine Guns, 7.62mm, M60."

The *Department of Defense Index of Specifications and Standards (DODISS)* (D7.14:(yr./pts.)) is an important source of information on unclassified federal, military, and departmental specifications and standards as well as those industry standards that are adopted by the DoD. *DODISS* is issued annually (with bimonthly supplements) and is available on a subscription basis from the Superintendent of Documents, U.S. Government Printing Office, Washington, DC 20402. A bimonthly microfiche edition of *DODISS* is published also. Each bimonthly microfiche edition completely supersedes the previous microfiche edition. Industries and individuals can obtain the microfiche edition of *DODISS* on an annual subscription basis from:

Navy Publications and Printing Service Office
Fourth Naval District
700 Robbins Avenue
Philadelphia, PA 19111

New standards and specifications and amendments are also announced in journals and technical magazines such as *Technical Communication: Journal of the Society for Technical Communication.*

The following are additional sources of information on military standards and specifications:

Naval Ordnance Systems Command. *Index of Ordnance Specifications and Weapons Specifications.* Issued annually, with midyear supplements. Available from: The Commanding Officer, Naval Ordnance Station, Central Technical Documents Office, Louisville, KY 40214, Attn: Code 80121.

Government Specifications and Standards for Plastics Covering Defense Engineering Materials and Applications (Revised Final). Norman E. Beach (Plastics Technical Evaluation Center, Picatinny Arsenal, Dover, NJ, May 1973). Available from NTIS as AD 771 008 (D4.11/2:6C).

Master Index of Government Guide Specifications for Construction, 3rd edition. Edited by Donald Bosserman (District of Columbia Metropolitan Chapter, Construction Specification Institute, 567 Southlawn Lane, Rockville, MD 20850), 1981, 123p.

Copies of unclassified military specifications and standards listed in the *DODISS* may be obtained from the DoD Single Stock Point (DOD-SSP) located in the Naval Publications and Forms Center (NPFC), 5801 Tabor Avenue, Philadelphia, PA 19120. Besides military standards and specifications, the DOD-SSP also distributes industry standards, federal standards and specifications, military handbooks, and other government standards documents.

A booklet entitled *Department of Defense Single Stock Point for Specifications and Standards: A Guide for Private Industry* (available from the Naval Publications and Forms Center, 5801 Tabor Avenue, Philadelphia, PA 19120) describes the DOD-SSP and provides information on the acquisition of military standards and specifications, which are required by those doing business with the DoD. Requests for standards and specifications should preferably be sent on DD Form 1425; telephone requests may also be placed by calling (215)697-3321.

Various standardization functions have been assigned to 68 offices in the DoD: the addresses and scopes of interest of these offices are listed in *The Standardization Directory* (SD-1, 1977, D7.13:1/23), which is available from the Naval Publications and Forms Center, 5801 Tabor Avenue, Philadelphia, PA 19120.

The DoD cooperates with other government agencies and private sector organizations in their standardization activities. The *Defense Standardization Manual* (D7.6/3:4120.3-M) contains a statement of procedures for adopting industry standards to avoid duplication of standardization effort. Industry standards adopted by the DoD are also listed in the *DODISS.*

CONSUMER PRODUCT SAFETY COMMISSION (CPSC)

The CPSC is an independent federal regulatory agency, established in May 1973 to implement the Consumer Product Safety Act (P.L. 92-573). The primary goal of the commission is to reduce injuries associated with consumer products. The commission also administers four other acts previously handled by other agencies: the Flammable Fabrics Act, the Federal Hazardous Substances Act, the Poison Prevention Packaging Act, and the Refrigerator Door Safety Act. The objectives of the commission are

to protect the public against unreasonable risks of injury from consumer products;

to assist consumers to evaluate the comparative safety of consumer products;

to develop mandatory safety standards for consumer products and minimize conflicting state and local requirements; and

to promote research and investigation into the causes and prevention of product-related deaths, illnesses, and injuries.

The commission issues standards independently and also encourages development of product safety standards by other agencies. The Office of Standards Coordination and Appraisal is responsible for the development of standards; the Bureau of Compliance enforces the standards. Standards issued by the CPSC are published in the *Federal Register*. Final regulations are published in Title 16, Chapter 2, of the *Code of Federal Regulations*. A complete list of standards issued by the commission can be obtained from the Office of Standards Coordination and Appraisal, Consumer Product Safety Commission, 1111 Eighteenth Street, NW, Washington, DC 20207.

OTHER SOURCES OF FEDERAL STANDARDS AND SPECIFICATIONS

As mentioned before, a great number of federal government agencies develop and use standards and specifications. Particularly notable are agencies such as the Environmental Protection Agency, the National Aeronautics and Space Administration, the General Services Administration, and the Departments of Housing and Urban Development, Labor, Energy, Transportation, and Agriculture, to name a few. Federal standards and specifications are administered by the General Services Administration. Federal specifications are designated by numbers with three components indicating: 1) the federal procurement group to which the item belongs, 2) the first letter of the specification title, and 3) a serial number; for example,

L-P-349C "Plastic Molding and Extrusion Material, Cellulose Accetate Butyrate."

In this designator, the first letter L represents the federal procurement group Cellulose Products and Synthetic Resins; the letter P is the first letter of the title of the specification; and the third component is a serial number. Similarly, federal standards have designators consisting of three components:

FED-STD-601 "Rubber: Sampling and Testing."

Federal standards and specifications are indexed in the *Index of Federal Specifications Standards and Commercial Item Descriptions* (GS2.8/2:(date)), published annually since 1952 by the General Services Administration, with monthly supplements. Copies of federal standards and specifications can be obtained from

General Services Administration
Specifications and Consumer Information Distribution Section
Washington Navy Yard, Building 197
Washington, DC 20407

or

The Department of Defense Single Stock Point
Naval Publications and Forms Center
5801 Tabor Avenue
Philadelphia, PA 19120

Brief descriptions of the standardization activities of hundreds of organizations, including fifty federal government agencies, are given in the *Directory of United States Standardization Activities*, edited by Sophie J. Chumas (National Bureau of Standards Special Publication 417) (Washington, DC: National Bureau of Standards, 1975) (C13.10:417).

REFERENCES

1. *ISO Definitions 1. Standardization Vocabulary: Basic Terms and Definitions* (Geneva: International Organization for Standardization, 1971).
2. A. S. Tayal, "Acquisition and Updating of Standards and Specifications in Technical Libraries," *UNESCO Bulletin for Libraries* 25 (July 1971): 198-204.
3. S. K. Reeves, "Specifications, Standards, and Allied Publications for U.K. Military Aircraft," *ASLIB Proceedings* 22 (September 1970): 432-48.
4. Bernard Houghton, *Technical Information Sources*, 2d edition (Hamden, CT: Linnet Books, Shoe String Press, 1972), pp. 67-68.
5. Denis J. Grogan, *Science and Technology: An Introduction to the Literature*, 3d edition (London: Clive Bingley, 1976), pp. 275-76.
6. David Hemenway, *Industrywide Voluntary Product Standards* (Cambridge, MA: Ballinger Publishing Co., 1975), pp. 8-9.
7. *Standards and Certification: Proposed Rule and Staff Report* (Washington, DC: Federal Trade Commission, Bureau of Consumer Protection, December 1978), pp. 31-38.

8. Erasmus J. Struglia, *Standards and Specifications Information Sources* (Management Information Guide 6) (Detroit, MI: Gale Research Company, 1965).

9. K. Subramanyam, *Scientific and Technical Information Resources* (New York: Marcel Dekker, 1981).

10. Rexmond C. Cochrane, *Measures for Progress: A History of the National Bureau of Standards* (Washington, DC: National Bureau of Standards, 1966) (C13.10:275).

11. Herman M. Weisman, "Technical Librarians and the National Standard Reference Data System," *Special Libraries* 63 (February 1972): 69-76.

FURTHER READING

Chumas, Sophie J., ed. *Directory of United States Standardization Activities* (NBS Special Publication 417). Washington, DC: National Bureau of Standards, 1975 (C13.10:417).

Henderson, M. M. "Standards: Development and Impact." *Special Libraries* 72 (April 1981): 142-48.

Kuiper, Barteld E. "Toward a World Catalog of Standards." *UNESCO Bulletin for Libraries* 27 (May 1973): 155-59+.

Slattery, William J. "Standards Information Service (NBS)." *Information Hotline* 8 (October 1976): 30-31.

Verman, Lal C. *Standardization: A New Discipline*. Hamden, CT: Archon Books, Shoe String Press, 1973.

9

AUDIOVISUAL AND NON-BOOK RESOURCES*

To ignore the audiovisual media produced by the government in the areas of science and technology is to ignore a vast and diverse information store. One of the characteristics of government-produced AV materials that makes them valuable is their diversity in subject matter and media format. They may deal with medicine, nutrition, or space exploration, just to name a few, and may be presented as films, posters, charts, maps, recordings, picture sets, videotapes, or photographs, in addition to other formats.

One of the reasons for this diversity is the fact that numerous agencies are involved in producing AV formats. In fiscal year 1980, for example, forty-four executive departments and agencies reported some type of audiovisual activity, including NASA, the National Science Foundation, and the Departments of Agriculture, Defense, Energy, and Interior. A total expenditure of $61,346,800 has been calculated for government audiovisual production for 1980 alone.[1]

With such a variety of agencies involved with science and technology, they can be important information resources for individuals in the sciences. Unfortunately, many of the people who could benefit from them have never used government-produced AV materials, simply because they were unaware that they existed.

*This chapter was first published in *RQ* as "U.S. Government Produced Audiovisual Materials," *RQ* 21 (Fall 1981): 23-26.

It is not uncommon for researchers to assume that the government's information production is limited to the more familiar print formats—books, journals, pamphlets and the like. This is not surprising, since non-print materials cannot be traced using the traditional indexes and bibliographic tools used for print formats. To locate AV materials, a separate set of searching tools must be used. When one becomes familiar with these special tools, locating AV material is not much more difficult than locating a monograph or journal produced by the government.

This chapter will focus upon strategies for identifying and obtaining government-produced AV materials. The first step in this process is simply finding out which AV materials exist in an area of interest. This step, which could be called bibliographic access, involves utilization of the appropriate indexes and lists to locate bibliographic citations for appropriate AV material. The second step, physical access, involves obtaining these materials for use. Many of the AV productions may be available through free loan, short-term rental, or purchase from, among others, federal agencies.

BIBLIOGRAPHIC ACCESS

Government production of AV materials is even more decentralized than its more traditional print output. The Government Printing Office (GPO), responsible for serving the printing needs of the Congress and the federal departments and agencies, functions as the official government printer for books, journals, pamphlets, and other print formats. Although a lot of government printing is done outside the GPO, the presence of this centralized printing facility provides some control for the efforts of the Superintendent of Documents in compiling catalogs and indexes to government publications.

In the case of government-produced AV formats, no such centralized production facility exists. Individual federal agencies and departments produce their own films and other AV productions, or contract with outside sources to produce them. The absence of a central production facility such as the GPO makes the task of identifying and recording the existence of these productions more difficult. As a result, no master list of all government AV materials has ever been compiled and maintained. In order to locate these materials, several sources must be consulted in an attempt to construct a complete list of available items. The information sources that offer the broadest, although not complete, coverage are those produced by the National Audiovisual Center (NAC).

NATIONAL AUDIOVISUAL CENTER (NAC)

The National Audiovisual Center serves as the central information clearinghouse and distribution center for U.S. government-produced audiovisual materials.[2] Since its creation in 1969, the NAC has aimed to make federally produced AV material available to the general public and to specialized users. This availability is achieved through attempts to compile bibliographic records of government-produced non-print material, as well as providing for physical access to these productions through the distribution programs of the NAC.

Before the NAC was established, federally produced AV materials had been obtainable for sale or loan largely by contacting individual federal agencies or their divisions and offices. Verifying the existence of materials on a particular topic was even more difficult, and often involved searching for lists of AV materials on an agency-by-agency basis. Today the NAC's centralized collections include more than 13,000 titles in numerous media formats, with new titles added daily. AV formats handled by the NAC include films, slide sets, filmstrips, videotapes, videocassettes, and multimedia kits produced by nearly three hundred federal agencies.

Many of these AV productions were designed for general use; others are aimed at specific training or instructional programs. Some of the science-related subject areas represented in the NAC collections include medicine, dentistry, and the allied health sciences; aviation and space technology; environmental sciences; atomic energy; and electronics. Occasionally these AV resources have been produced to be used in conjunction with print information, as in the case of a film and companion discussion leader's guide. In other cases, they are meant to stand alone.

Finding aids produced by the NAC include the publication titled *A Reference List of Audiovisual Materials Produced by the United States Government*, the update medium titled *Films, Etc.*, minicatalogs in specific subject areas, and brochures on specific topics or AV formats.

A Reference List of Audiovisual Materials Produced by the United States Government (GS4.2:Au2/978) is the most comprehensive of these finding aids and is the title that should be consulted first when trying to identify AV productions in science and technology as well as other subject areas. This source can be searched by subject areas or by specific titles of AV productions. A list of subject headings used is given in the front of the volume. The information provided for each AV item listed includes title, format, short description, availability for rent or sale, and price.

Although *A Reference List of Audiovisual Materials Produced by the United States Government* is the most comprehensive index published by the NAC, it does not always include the most up-to-date AV listings available. New editions of *A Reference List* have been published on an irregular basis, with supplements occasionally issued between editions. The NAC plans to issue quarterly updates to *A Reference List* in the future, and this will aid in the attempt to provide more current listings. Two sources that offer more current listings, but on a less comprehensive scale, are *Films, Etc.* and the subject minicatalogs.

Films, Etc. is a current awareness listing, distributed irregularly by NAC. This publication describes selected audiovisual productions in several subject areas and provides an annotation, bibliographic information, and sale or rental price for each item. Although *Films, Etc.* offers a broad overview, the subject minicatalogs focus on particular topics and are issued for several general subject areas in science and technology including "Engineering," "Medicine," and "The Sciences." In addition to *Films, Etc.* and the minicatalogs, the NAC issues brochures describing both individual titles and multiple titles, as well as listings of titles arranged by media format or by individual agency-producers. Any of these free materials may be requested individually from the NAC Reference Section.

The NAC Reference Section fields written and telephone requests for information about audiovisuals, makes referrals to free loan sources for some government films, and maintains a computerized master data file for government audiovisuals. The NAC Reference Section is also the source to contact for free information lists, catalogs, and special brochures that describe the NAC collection. Contact them through: National Audiovisual Center, National Archives and Records Service, GSA, Reference Section/PC, Washington, DC 20409; (301)763-1896.

The NAC has also published a useful source titled *Directory of U.S. Government Audiovisual Personnel* (GS4.24:977). This reference tool lists federal agencies and personnel involved in radio, television, motion pictures, photography, and sound recordings, along with their mailing addresses and telephone numbers.

THE NATIONAL LIBRARY OF MEDICINE (NLM)

Although the National Audiovisual Center attempts to provide centralized control for government-produced AV materials in all subject areas, the National Library of Medicine (NLM) has assumed this responsibility for the medical sciences. The medical sciences are among the most active producers and users of audiovisual formats, and NLM is the major source of bibliographic control for this array of materials.

Since 1978 NLM has listed audiovisuals in a separate publication titled *National Library of Medicine Audiovisuals Catalog* (HE20.3609/4:date). This quarterly publication allows hundreds of new non-print materials received and cataloged by NLM to be identified through either subject or name/title searches. Bibliographic information given for each item includes title and producer, medium, sale or loan source, price, and subject headings under which the item has been classified.

The *NLM Audiovisuals Catalog* focuses upon two general categories of AV material. For the first category, material which has been developed for instructional use in health science education, the catalog provides abstracts describing content. The second category of materials, recordings of educational events such as congresses, symposia, and lectures, are listed without abstracts.

The National Library of Medicine also maintains a computerized data base of AV information called AVLINE (Audiovisuals On-Line). This data base, which is used to compile the *NLM Audiovisuals Catalog*, is searchable online from terminals located at more than 800 U.S. institutions with access to the NLM online network. AVLINE includes more titles than the printed catalog because of certain omission policies for the print catalog and because of the currency of an online system. It also includes evaluative comments for included materials.

MONTHLY CATALOG OF U.S. GOVERNMENT PUBLICATIONS

Traditionally, the *Monthly Catalog* has not listed audiovisual productions themselves, but only print publications associated with them, such as user guides or filmographies. This should change in the future, however, as a result of a cooperative cataloging agreement between the Library of Congress, the GPO Library, and the National Audiovisual Center. As a result of this agreement, a catalog record for all material distributed by the NAC will appear in the *Monthly Catalog*.[3] This change will make it easier to identify pertinent AV productions while searching a specific topic in the *Monthly Catalog*.

INDIVIDUAL AGENCY LISTS

Many agencies publish separate lists of their own audiovisual productions. Often these agency lists are distributed through the National Audiovisual Center. At other times they are distributed through the agency or through the Superintendent of Documents sales program. There are several approaches to identifying individual agency lists. The first, and simplest is to have one's name added to the National Audiovisual Center mailing list. The National Audiovisual Center automatically mails AV lists appropriate to the user's subject interest profile.

A second approach is to search the *Monthly Catalog* under agency name or appropriate subject headings and keywords. Although the *Monthly Catalog* does not currently list audiovisual publications, it does index many print publications associated with them, such as bibliographies, filmographies, and user guides.

A third method for acquiring agency AV lists is to request them directly from the agency. Agency addresses and telephone numbers can be located using the *U.S. Government Manual*, available in almost any library.

THE U.S. GEOLOGICAL SURVEY

Since 1879 the U.S. Geological Survey has been engaged in topographic and geologic mapping and collecting information about public lands. The survey's National Mapping Program has since been established to provide multipurpose maps and related data, with emphasis on topographic maps. A topographic map is a line-and-symbol representation of natural and man-made features on a section of the earth's surface, plotted to a definite scale. The National Mapping Program also produces special-purpose maps and related map data, including orthophotomaps; orthophotoquads; county maps; quadrangle maps; state base maps; maps of national parks, monuments, and historic sites; Antarctic maps; and maps of the entire United States.

THE NATIONAL OCEAN SURVEY (NOS)

The National Ocean Survey (NOS), begun in 1807, is the federal government's oldest scientific agency. For more than 170 years NOS has collected and published data about the oceans and land, including maps and charts based on hydrographic and topographic surveys.

NOS nautical charts are available in various scales for the United States coastline, and that of its territories and possessions, and the Great Lakes. Specialized products such as bathymetric maps and storm evacuation maps are also available from NOS.

NOS aerial photographs are available for coastal areas and most civil airfields. These may be purchased in black and white, infrared, and natural color photographs.

NOS technical exhibits are available for meetings, conventions, and other scientific gatherings. Comprised of self-standing wall panels, the exhibits depict the history of the agency and current survey techniques and activities.

NATIONAL CARTOGRAPHIC INFORMATION CENTER (NCIC)

The National Cartographic Information Center (NCIC) is a unit of the U.S. Geological Survey that exists to help the public find maps of all kinds, as well as much of the data and materials used to compile and print them. It is the public's primary source of cartographic information. The NCIC deals with not just federal, but also state, local, and private cartographic information. The NCIC sells map data products, provides research and information services, and provides information about numerous federal agencies engaged in mapping.

The NCIC operates five regional offices and has established affiliated offices with many state governments. The address of NCIC Headquarters is National Headquarters, National Cartographic Information Center, U.S. Geological Survey, 507 National Center, Reston, VA 22092; (703)860-6045.

FEDERAL AGENCY MAPPING ACTIVITIES

The listing below identifies the types of cartographic products produced by specific federal agencies. To contact these agencies for further information, consult the *U.S. Government Manual* for their addresses and telephone numbers. Most agencies will send catalogs of their map products upon request. This is an excellent way to find out what is available.

U.S. Forest Service: Aerial photos, national forest maps, and recreation maps.

Bureau of Land Management: Cadastral surveys, aerial photos, federal land maps, and land-use maps.

Water and Power Resources Service: (Formerly Bureau of Reclamation.) Aerial photos and river surveys.

U.S. Geological Survey: Aerial photos, topographic maps, orthophotoquads, county maps, regional maps, state maps, United States maps, polar region maps, satellite image maps, lunar and planetary maps, geophysical maps and charts, geologic maps, federal water resource development maps, bathymetric maps, hydrologic and flood-related maps, river survey maps, mineral and energy resources maps, the National Atlas, geodetic control lists and diagrams, map and aerial photo certifications, lists and gazetteers of geographic names, open file reports, flood-prone area maps, land-use maps, boundary information, and color separation materials.

Bureau of the Census: Demographic maps, SMSA maps, and congressional district maps.

Central Intelligence Agency: Country and world maps.

National Oceanic and Atmospheric Administration and its **National Ocean Survey:** Aerial photos, planimetric maps, nautical charts, aeronautical charts, bathymetric maps, historical maps, geodetic controls, flood evacuation maps, topographic maps, coastal zone maps, and climate maps.

Corps of Engineers: River navigation charts, topographic maps, and geodetic surveys.

Federal Highway Administration: Transportation maps and county highway maps.

Federal Power Commission: Utility maps.

Tennessee Valley Authority: Aerial photos, topographic maps, bathymetric maps, recreation maps, nautical charts, utility maps, and geodetic surveys.

Mississippi River Commission: Topographic maps and river charts.

International Boundary Commission: Maps and international boundaries.

Library of Congress: Maps, charts, atlases, and globes.

Agricultural Stabilization and Conservation Service: Aerial photos.

Soil Conservation Service: Aerial photos, Landsat image maps, and soil survey maps.

National Archives and Records Service: Aerial photos, maps, and charts.

National Aeronautics and Space Administration: Aerial photos, space photos, atlases of planets, and space exploration photos.

Defense Mapping Agency: Aerial photos, topographic maps, hydrologic maps, nautical maps, aeronautical charts, lunar and planetary maps, military installation maps, and digital terrain tapes.

INDEXES OF MAP DATA

Special indexes that show every topographic map published for each state and for U.S. island territories are free upon request from the addresses given below:

Maps of areas east of the Mississippi:

Branch of Distribution
U.S. Geological Survey
1200 South Eads Street
Arlington, VA 22202

West of the Mississippi:

Branch of Distribution
U.S. Geological Survey
Box 25286, Federal Center
Denver, CO 80225

Residents of Alaska may write:

Distribution Section
U.S. Geological Survey
Federal Bldg., Box 12
101 Twelfth Avenue
Fairbanks, AK 99701

These indexes contain order forms for ordering maps, addresses of local map reference libraries, local map dealers, and federal map distribution centers.

Several bibliographic indexes and abstracts include references to maps along with other publications. These indexes are available in print or machine-readable formats, sometimes in both. Government-produced indexes that cite maps include:

Government Reports Announcements & Index and the NTIS Bibliographic Data File, both from the National Technical Information Service

GPO *Monthly Catalog* (print and computer data base formats)

Publications of the U.S. Geological Survey (an internal data base for staff only)

SUBJECT BIBLIOGRAPHIES

The Superintendent of Documents publishes a series of free bibliographies of government materials available on particular topics. This series, called Subject Bibliographies, includes two titles related to government AV materials. The subject bibliography titled "Posters, Charts, Picture Sets and Decals" (SB-057) lists materials in these formats available on various topics. "Motion Pictures, Films, and Audiovisual Information" (SB-073) lists sources of print information about audiovisual materials. Both provide price, ordering information, and Superintendent of Documents classification number for each item listed. To request free copies of these subject bibliographies, write: Superintendent of Documents, U.S. Government Printing Office, Washington, DC 20402.

PHYSICAL ACCESS

NATIONAL AUDIOVISUAL CENTER (NAC)

The NAC is the major distributor of government audiovisual materials. Any of the AV materials listed in *A Reference List of Audiovisual Materials Produced by the United States Government*, in *Films, Etc.*, in the subject minicatalogs, or in NAC-distributed brochures can be obtained through NAC either for purchase, rental, and occasionally for preview. The information sources above will provide information on sale or rental price as well as preview availability.

The NAC will also supply copies of any supplementary materials, such as study guides or teacher manuals, produced to accompany specific audiovisual productions. In addition, the center can convert its 16mm films to videocassette formats. To receive a price quote, supply them with film title and make and model of the videotape player to be used.

NATIONAL LIBRARY OF MEDICINE (NLM)

On the other hand, materials listed in the *NLM Audiovisuals Catalog* are not distributed through NLM. Instead, each listing in the catalog indicates the source for purchase or loan of the item. At the back of each catalog full addresses and telephone numbers are given for these distributors.

OTHER AGENCIES

Free loans are available for many of the government-produced AV materials. Often a free loan can be arranged by contacting the producing agency or one of its field offices directly. The producing agency may provide a reference to a commercial film distributor who has agreed to distribute certain agency films without charge. Although the National Audiovisual Center does not lend materials, their reference section will refer a person to federal agency lending libraries and commercial loan sources for government-produced films.

CARTOGRAPHIC INFORMATION

Like other government-produced materials, maps may be purchased or borrowed from a library that collects them. Some libraries circulate maps; others restrict their use to within the library. Many libraries have built up good map collections as a result of map deposit collection arrangements with agencies such as the U.S. Geological Survey and the Defense Mapping Agency. Federal depository libraries also receive maps on deposit.

Cartographic products may be purchased in numerous formats, including photographs, slides, microfilm, and machine-readable formats. To purchase these materials, the map-producing agency may be contacted directly. Addresses of such agencies may be obtained from *U.S. Government Manual.*

Another source of map and map purchase data is the National Cartographic Information Center. NCIC can find out what data are available, identify the agency holding the data, and give instructions for ordering. The address of NCIC has been given elsewhere in this chapter.

REFERENCES

1. U.S. National Audiovisual Center, *Federal Audiovisual Activity, Fiscal Year 1980* (Washington, DC: The Center, 1981), p. 3.

2. Margery J. McCauley, "Information Policy and the National Audiovisual Center," *Government Publications Review* 8A (1981): 215-20.

3. "Agreement Set on Audiovisual Cataloging," *LC Information Bulletin* 40 (November 1981): 389-90.

FURTHER READING

Executive Office of the President. Office of Management and Budget. Office of Federal Procurement Policy. *Federal System for Acquiring Audiovisual Productions* (OFPP Pam. No. 3). Washington, DC: Office of Management and Budget, 1980 (PrEx2.22:3).

Medical Library Association. *Index to Audiovisual Serials in the Health Sciences: A Publication of the Medical Library Association Produced in Cooperation with the National Library of Medicine.* Chicago: The Association, 1977- . Quarterly.

National Aeronautics and Space Administration. *NASA Films.* Washington, DC: NASA, 1981 (NAS1.2:F48/981).

National Audiovisual Center. *Directory of U.S. Government Audiovisual Personnel.* 6th ed. Washington, DC: NAC, 1977 (GS4.24:977).

National Audiovisual Center. *A List of Audiovisual Materials Produced by the United States Government for Emergency Medical Service.* Washington, DC: NAC, 1980 (GS4.17/5-2:M46).

National Audiovisual Center. *A List of Audiovisual Materials Produced by the United States Government for Environment and Energy Conservation.* Washington, DC: NAC, 1980 (GS4.17/5-2:En8).

National Audiovisual Center. *A List of Audiovisual Materials Produced by the United States Government for Fire/Law Enforcement.* Washington, DC: NAC, 1980 (GS4.17/5-2:F51).

National Audiovisual Center. *A List of Audiovisual Materials Produced by the United States Government for Flight and Meteorology.* Washington, DC: NAC, 1980 (GS4.17/5-2:F64).

National Audiovisual Center. *A List of Audiovisual Materials Produced by the United States Government for Nursing.* Washington, DC: NAC, 1980 (GS4.17/5-2:N93).

National Audiovisual Center. *Medical Catalog of Selected Audiovisual Materials Produced by the United States Government.* Washington, DC: U.S. Government Printing Office, 1980 (GS4.17/6:980).

National Audiovisual Center. *A Reference List of Audiovisual Materials Produced by the United States Government.* Washington, DC: NAC, 1978 (GS4.2:Au2/978).

National Audiovisual Center. *A Reference List of Audiovisual Materials Produced by the United States Government. Supplement.* Washington, DC: U.S. Government Printing Office, 1980 (GS4.2:Au2/980/supp.).

National High Blood Pressure Education Program. *Audiovisual Aids for High Blood Pressure Education.* (DHEW Pub. No. NIH 80-1663). Bethesda, MD: National Heart, Lung, and Blood Institute, 1979 (HE20.3202:Au2).

National Institute on Drug Abuse. Resource Center. *Audiovisual Catalog—NIDA Resource Center.* Rockville, MD: The Institute, 1980 (HE20.8211: Au2).

National Library of Medicine. *National Library of Medicine Audiovisuals Catalog.* Washington, DC: U.S. Government Printing Office, 1977- . Quarterly (HE20.3609/4:date).

National Medical Audiovisual Center. *National Medical Audiovisual Center Catalog: Films for the Health Sciences.* Washington, DC: U.S. Government Printing Office, 1981 (HE20.3608/4:981).

Project Media Base. *Problems in Bibliographic Access to Non-Print Materials: Final Report.* Washington, DC: U.S. Government Printing Office, 1979 (Y3.L61:2B47).

U.S. Department of Agriculture. Science and Education Administration. *Beef Cattle Publications and Visual Materials.* (Misc. Pub. No. 1398). Washington, DC: U.S. Government Printing Office, 1980 (A1.38:1398).

U.S. Department of Agriculture. Science and Education Administration. *Horse Publications and Visual Materials.* (Misc. Pub. No. 1393). Washington, DC: U.S. Government Printing Office, 1980 (A1.38:1393).

U.S. Department of Agriculture. Science and Education Administration. *Sheep Publications and Visual Materials.* (Misc. Pub. No. 1399). Washington, DC: U.S. Government Printing Office, 1980 (A1.38:1399).

U.S. Department of Agriculture. Science and Education Administration. *Swine Publications and Visual Materials.* (Misc. Pub. No. 1397). Washington, DC: U.S. Government Printing Office, 1980 (A1.38:1397).

U.S. Food and Drug Administration. *Catalogue of FDA Publications and Audiovisual Materials for Consumers.* (DHEW Pub. No. FDA 77-1030). Rockville, MD: FDA, 1977 (HE20.4016:P96).

U.S. Forest Service. Eastern Region. *Forest Service Films Available on Loan to the Public for Educational Purposes.* Milwaukee, WI: Forest Service, Eastern Region, 1979 (A13.2:F48).

U.S. Forest Service. Pacific Northwest Region. *Film Catalog: Pacific Northwest Region.* Portland, OR: Forest Service, Pacific Northwest Region, 1980 (A13.2:F48/2/1980).

U.S. Forest Service. Southern Region. *Forest Service Films 1979, Southern Region.* Atlanta, GA: Forest Service, Southern Region, 1979 (A13.2: F48/3).

U.S. Superintendent of Documents. *Motion Pictures, Films, and Audiovisual Information.* (Subject Bibliography 073). Washington, DC: U.S. Government Printing Office, 1980 (GP3.22/2:073/6).

U.S. Superintendent of Documents. *Posters, Charts, Picture Sets, and Decals.* (SB 057). Washington, DC: U.S. Government Printing Office, 1980 (GP3.22/2:057/7).

10

INDEXES AND ABSTRACTS

The previous chapters of this guide concentrated on various types of primary literature published by the federal government, which included technical reports, journal articles, patents, standards and specifications, translations, and non-print materials. Generally, primary literature is scattered and unorganized. For example, technical reports on a given subject may be issued by a number of different federal agencies and federal contractors. A number of different periodicals may carry articles on a given topic.

Consequently, to be able to retrieve this primary literature quickly and effectively, it is necessary to organize this material and to bring it under bibliographic control by means of secondary literature services such as abstracts and indexes. Abstracting and indexing services, by definition, do not directly contribute to the scientific knowledge but provide access to the primary literature by means of appropriate retrieval keys such as authors' names, topical headings, report and patent numbers, and corporate authors.

The federal government issues a great number of indexing and abstracting services in the science and technology areas. Some of these services, which have already been described, deal with the federally produced primary literature. For example, *Monthly Catalog of United States Government Publications, Government Reports Announcements & Index, Technical Abstracts Bulletin*, and *Official Gazette of the United States Patent and Trademark Office* attempt to bring the government publications, reports, and patents under bibliographic control. Although some of these sources do not achieve complete coverage, they are nevertheless the most comprehensive sources available within their areas of responsibility.

Another type of secondary services is the publications catalogs of various agencies. Clearly, these catalogs are of much narrower scope but their purpose still is to provide effective access to the respective agencies' publications. The *Publications Catalog of the U.S. Department of Health and Human Services* and the *Publications of the National Bureau of Standards* are examples of this type of secondary literature.

Equally important, however, are the abstracting and indexing services published and/or sponsored by the federal government whose scope goes beyond the federal publications. Examples of such services are *Index Medicus, Energy Research Abstracts*, and *Selected Water Resources Abstracts*. Typically, such services are initiated by the federal agencies to meet the information needs of their scientists. These are mission-oriented and attempt to cover journal literature, monographs, theses, and conference papers so that their researchers have access to current information published in their fields.

Starting of such services is justified on the basis that the existing abstracting and indexing services do not adequately serve the needs of the scientists who are engaged in a specific mission. The federal involvement in such products is extensive. For example, the most famous of the federal indexing services, *Index Medicus*, has been in existence since 1879.[1] Geological Survey started the compilation of a bibliography in 1883 which later became the *Bibliography of the Geology of North America* which, in turn, was succeeded by the *Bibliography and Index of Geology* published by the American Geological Institute. *Nuclear Science Abstracts*, during its publication, was the preeminent service in its field.

Another aspect that has to be kept in mind is the federal assistance to many abstracting and indexing services produced by the non-governmental organizations. Adkinson traces the cooperation between federal agencies and outside organizations by citing the examples of *Bibliography and Index of Geology, Meteorological Abstracts*, and *Chemical Abstracts*.[2] Adkinson concludes that this relationship

> ... was an important factor in enabling abstracting and indexing services in the following fields to achieve or maintain preeminence among the world's A&I services: chemistry, astronomy, mathematics, earth sciences, biology, meterorology, aerospace, metals, engineering, psychology, physics, and many specialized fields.[3]

This cooperation also resulted in the increasing use of computerization in the production of abstracting and indexing services in general. For example, as early as 1949, the Atomic Energy Commission made use of punched cards for preparing indexes. Similarly, the Office of Science and Information Service of the National Science Foundation assisted many abstracting and indexing services in the computerization of their systems.[4] The computerization, of course, led to the development of several federal data bases many of which are described in chapter 11 on data bases.

The discipline or mission-oriented abstracting and indexing services published by the government or even the cooperative ventures such as those mentioned above are subject to the vagaries of the federal policies and priorities. Consequently, the abstracting and indexing services are subjected to the cyclical process where new services are continued to be added while others are discontinued. Adding to this are the publication policies of the recent federal administrations which tend to shift the publication responsibilities for some services to the private sector. *Bibliography of Agriculture*, for example, represents such a trend where a former federal publication is taken over by a commercial publisher.

In any event, the federal government's involvement in the publication of secondary services in the area of science and technology is significant and, in some areas such as aeronautics, medicine, and energy, the federal publications are dominant. Consequently, researchers and information specialists need to be aware of the federal abstracting and indexing services.

A promising source in locating federally published abstracting and indexing services is *Abstracting and Indexing Services Directory* (Detroit: Gale Research Company, 1982-). The directory, according to its subtitle is "a descriptive guide to abstracting journals, indexes, digests, serial bibliographies, catalogs, title announcement bulletins, and similar information access and alerting publications in all areas of science, technology, medicine, business, law, social sciences, education, and humanities." The first issue of the first edition of the directory is dated July 1982 and is expected to be issued in 6-month intervals. Each issue is expected to contain about five hundred listings which do not duplicate previous listings. The scope of the directory includes the publications of the government agencies.

The information provided on each abstracting and indexing service listed includes: title, publisher's name, address and telephone number, starting date, editor, description and scope, subject coverage, sources scanned, contents and arrangement, description of a typical entry, frequency/cumulations, subscription information, former titles, document delivery, computer access, microform availability, and other information.

INDEX AND ABSTRACT LISTINGS

The remainder of this chapter provides a listing of selected government-produced indexes and abstracts. Information is given on the scope, frequency, availability, and government producer of each. In cases where a print index is complemented by a companion machine-readable data base, the name of the data base is provided.

Title: *Abridged Index Medicus*, 1970- .
Producer: National Library of Medicine
Scope: Indexes 118 English-language biomedical journals each month, with the mission of serving the needs of individual practitioners, and small hospital and clinical libraries. Its more comprehensive counterpart is *Index Medicus*.
Frequency: Monthly; an annual compilation is sold separately.
Availability: On subscription from the Superintendent of Documents; available to depository libraries (HE20.3612/2:(v.nos.&nos.))
Companion Resources: MEDLARS (Medical Literature Analysis and Retrieval System of the National Library of Medicine); *Medical Subject Headings* (*MeSH*)

Title: Abstracts of ARI Research Publications[5]
Producer: U.S. Army, Research Institute for the Behavioral and Social Sciences
Scope: Abstracts of Research Reports, Technical Papers, Utilization Reports, and Technical Reports published by the Army Research Institute, along with descriptions of intra-agency Research Problem Reviews and Research Memoranda; Abstracts for 1977 were published in 1980.
Availability: U.S. Army Research Institute for the Behavioral and Social Sciences, ATTN: PERI-TP, 5001 Eisenhower Avenue, Alexandria, VA 22333. Available to depository libraries (D101.60/4:).

Title: Aeronautical Engineering: A Continuing Bibliography, 1970- .
Producer: Scientific and Technical Information Branch, National Aeronautics and Space Administration
Scope: Selected, annotated references to unclassified reports and journal articles added to the NASA scientific and technical information system and announced in *Scientific and Technical Aerospace Reports* (*STAR*) and *International Aerospace Abstracts.* Subjects covered relate to engineering, design, and operation of aircraft and aircraft components.
Frequency: Monthly, with annual cumulative index.
Availability: National Technical Information Service. Also available to depository libraries (NAS1.21:7037(nos.)).

Title: Aerospace Medicine and Biology: A Continuing Bibliography, 1964- .
Producer: Scientific and Technical Information Branch, National Aeronautics and Space Administration and the American Institute of Aeronautics and Astronautics
Scope: An abstracting and announcement journal for references in bioscience and biotechnology. Selected, annotated references to unclassified reports and journal articles added to NASA's scientific and technical information system and announced in *Scientific and Technical Aerospace Reports* (*STAR*) and *International Aerospace Abstracts.*
Frequency: Monthly, with annual cumulative index
Availability: National Technical Information Service. Also available to depository libraries (NAS1.21:7011(nos.)).

Title: Annual Index of Rheumatology, 1965- .
Producer: Arthritis Foundation and National Library of Medicine
Scope: Includes references from world literature on rheumatology cited in *Cumulative Index Medicus*
Frequency: Annual
Availability: Arthritis Foundation, Finance Dept., 3400 Peachtree Road, NE, Atlanta, GA 30326; (404)266-0795. Prepayment is required.
Companion Resource: National Library of Medicine Data Base

Title: Bibliography of Agriculture, 1942- .
Producer: The Oryx Press, 2214 North Central at Encanto, Phoenix, AZ 85004
Scope: Produced from the computerized bibliographic data files distributed by the National Agricultural Library, it covers journal articles, pamphlets, government documents and reports in the areas of agriculture and allied sciences.
Frequency: Monthly
Availability: Oryx Press
Companion Resources: AGRICOLA (AGRICultural On-Line Access) data base; *National Agricultural Library Catalog* (Rowman and Littlefield). It is available online on BRS, DIALOG, and ORBIT.

Title: Cancergrams, 1977- .
Producer: International Cancer Research Data Bank, National Cancer Institute, National Institutes of Health
Scope: Current awareness bulletins with abstracts of recent research literature related to specific research topics. There are 65 different *Cancergrams*, each with monthly issues containing 30-100 abstracts pertinent to a specific cancer research topic. The abstracted papers are gleaned from over 3,000 biomedical journals and other sources.
Frequency: Monthly
Availability: Free to qualified researchers from the Cancer Information Dissemination and Analysis Center for Carcinogenesis Information, Science Information Services, Franklin Institute Research Laboratories, Twentieth and Race Streets, Philadelphia, PA 19103. Can be purchased from the National Technical Information Service. Available to depository libraries (HE20.3173/2:).

Title: Carcinogenesis Abstracts, 1963-1976.
Producer: National Cancer Institute, National Institutes of Health
Scope: Abstracts of significant articles on cancer from the world biomedical literature
Frequency: Monthly; ceased in 1976
Availability: Was available to depository libraries (HE20.3159: (v.nos.&nos.))

Title: Computer Program Abstracts, 1969-1981.
Producer: National Aeronautics and Space Administration
Scope: Abstracts of documented computer programs developed by or for NASA and the Department of Defense
Frequency: Quarterly; ceased in 1981
Availability: Free to NASA and affiliates from NASA's Scientific and Technical Information Facility, P.O. Box 8757, Baltimore/ Washington International Airport, MD 21240 (NAS1.44: (v.nos.&nos.))
Companion Resources: NASA/RECON (NASA's data base); *NASA Thesaurus*

Title: Dental Literature Index
Producer: American Dental Association, in cooperation with the National Library of Medicine
Scope: Listing of journal articles related to dentistry, produced using MEDLARS
Frequency: Quarterly, with annual cumulation
Availability: American Dental Association, Subscription Dept., 211 East Chicago Avenue, Chicago, IL 60611

Title: Diabetes Literature Index, 1966-1979.
Producer: National Institute of Arthritis, Metabolism, and Digestive Diseases, National Institutes of Health
Scope: Current scientific papers, worldwide, relevant to laboratory and clinical research into the nature of, causes, and therapy for diabetes. Prepared from the National Library of Medicine's MEDLARS data base.
Frequency: Monthly; ceased in 1979
Availability: Was free to National Institutes of Health grantees and contractors working in diabetes, and to medical school libraries, and was available to depository libraries (HE20.3310:(v.nos. &nos.)).
Companion Resources: MEDLARS (Medical Literature Analysis and Retrieval System of the National Library of Medicine); *Medical Subject Headings (MeSH)*

Title: EIA Data Index: An Abstract Journal, 1980- .
Producer: Energy Information Administration, U.S. Department of Energy
Scope: Indexes and abstracts individual tables, graphs, and other formatted data reported in publications of the Energy Information Administration. Subject areas of statistics include energy production, consumption, price, resource availability, and projections of supply and demand.
Frequency: Semiannual
Availability: For sale from the Superintendent of Documents. Available to depository libraries (E3.27/5:).
Companion Resources: The Federal Energy Data Index (FEDEX); *EIA Publications Directory: A User's Guide* (E3.27:980)

Title: EIA Publications Directory: A User's Guide, 1980- .
Producer: Energy Information Administration, U.S. Department of Energy
Scope: Abstracts of Energy Information Administration publications
Frequency: Quarterly
Availability: For sale from the Superintendent of Documents; available to depository libraries (E3.27:(yr.))
Companion Resources: The Federal Energy Data Index (FEDEX); *EIA Data Index: An Abstract Journal*

Title: EPA Publications Bibliography: Quarterly Abstract Bulletin, 1977- .
Producer: U.S. Environmental Protection Agency
Scope: Indexes EPA technical reports and journal articles added to the collections of the National Technical Information Service during the preceding quarter. The fourth issue of the year contains bibliographic citations with abstracts for the preceding quarter, and cumulative indexes for the calendar year.
Frequency: Quarterly
Availability: On subscription from the National Technical Information Service; available to depository libraries (EP1.21/7:(date))
Companion Resource: EPA Cumulative Bibliography which covered 1970-1976

Title: Earth Resources: A Continuing Bibliography with Indexes
Producer: Scientific and Technical Information Branch, National Aeronautics and Space Administration
Scope: A selection of annotated references to unclassified reports and journal articles related to remote sensing of earth resources by aircraft and spacecraft that were added to the NASA scientific and technical information system and announced in *Scientific and Technical Aerospace Reports* (*STAR*) and *International Aerospace Abstracts*
Frequency: Quarterly
Availability: National Technical Information Service. Available to depository libraries (NAS1.21:7041(nos.)).

Title: Endocrinology Index, 1968-1979.
Producer: National Institute of Arthritis, Metabolism, and Digestive Diseases, National Institutes of Health
Scope: Was a current awareness bibliography produced from the National Library of Medicine's MEDLARS data base. Aimed at scientists working in the field of endocrinology, to aid in integration of research and clinical efforts in endocrinology.
Frequency: Bimonthly; ceased in 1979
Availability: Was free to National Institutes of Health grantees and contractors working in endocrinology, and to medical school libraries, and was available to depository libraries (HE20.3309:(v.nos.&nos.))
Companion Resources: MEDLARS (Medical Literature Analysis and Retrieval System of the National Library of Medicine); *Medical Subject Headings* (*MeSH*)

Title: Energy: A Continuing Bibliography with Indexes, 1974- .
Producer: Scientific and Technical Information Branch, National Aeronautics and Space Administration
Scope: A selection of annotated references to unclassified reports and journal articles related to energy sources, solar energy, energy conversion, transport, and storage that were added to the NASA scientific and technical information system and announced in

Scientific and Technical Aerospace Reports (*STAR*) and *International Aerospace Abstracts*
Frequency: Quarterly
Availability: National Technical Information Service. Also available to depository libraries (NAS1.21:7043(nos.)).

Title: Energy Abstracts for Policy Analysis, 1975- .
Producer: Technical Information Center, U.S. Department of Energy
Scope: Focuses on non-technical and semi-technical documents on energy analysis and development, including programs; policy, legislation, and regulation; social, economic, and environmental impacts; regional and sectoral analyses. Abstracts congressional, federal agency, state, local, and regional reports; books, periodicals, and publications of industry and academia. (Technical energy documents are abstracted in *Energy Research Abstracts.*)
Frequency: Monthly, with annual cumulative index
Availability: Free to federal, state, and local agencies concerned with energy development; on subscription from the Superintendent of Documents; available to depository libraries (E1.11:(v.nos.&nos.))
Companion Resource: DOE Energy Data Base

Title: Energy Conservation Update
Producer: U.S. Department of Energy
Scope: Indexes and abstracts world literature on energy conservation. Topics range from transportation and industrial energy savings to research on heating, lighting, and hot water conservation.
Frequency: Monthly
Availability: On subscription from National Technical Information Service
Companion Resource: DOE Energy Data Base

Title: Energy Research Abstracts, 1977- .
Producer: Technical Information Center, U.S. Department of Energy
Scope: Subject areas covered include energy systems, conservation, safety, environmental protection, physical research, biology, medicine, reactive technology, radioactive waste processing and storage, nuclear fusion technology. Abstracts and indexes all technical reports, articles, conference papers and proceedings, books, patents, and theses of the DOE, its laboratories, energy centers, and contractors. Also abstracts other U.S. government-sponsored publications and international literature in energy areas.
Frequency: Semi-monthly, with annual and semiannual indexes
Availability: Free to federal, state, and local agencies dealing with energy development, conservation, and usage; on subscription from the Superintendent of Documents or from Technical Information Center, P.O. Box 62, Oakridge, TN 37830; available to depository libraries (E1.17:(v.nos.&nos.))
Companion Resource: DOE Energy Data Base

Title: FDA Clinical Experience Abstracts
Producer: Food and Drug Administration
Scope: Covers 180 U.S. and foreign biomedical journals. Human data on adverse effects, hazards, and efficacy of drugs, devices, and nutrients; adverse effects of cosmetics, household chemicals, pesticides and food additives; animal studies on teratogenicity and carcinogenicity.
Frequency: Monthly; ceased publication in 1976
Availability: Free to hospitals, physicians, pharmacists, public health officials, regulated industries and other individuals involved with safe use of drugs, cosmetics, etc.; available to depository libraries (HE20.4009:(v.nos.&nos.))
Companion Resource: Medical Subject Headings (MeSH)

Title: Federal Register Index
Producer: Office of the Federal Register, National Archives and Records Service
Scope: Indexes agency regulations published in the *Federal Register*
Frequency: Monthly, with annual cumulation in December issue
Availability: By subscription from the Superintendent of Documents; available to depository libraries (GS4.107:(v.nos.&nos.))

Title: Fire Technology Abstracts, 1976- .
Producer: Johns Hopkins University and the U.S. Fire Administration of the Federal Emergency Management Agency
Scope: Abstracts and indexes applied fire literature from books, journals, reports, patents, codes, and standards. Primarily English language, with selected entries from the world fire literature.
Frequency: Bimonthly, with annual cumulative index
Availability: On subscription from the Superintendent of Documents; available from depository libraries (FEM1.109:(v.nos.&nos.))

Title: Fossil Energy Update
Producer: U.S. Department of Energy
Scope: Abstracts and indexes current literature on fossil energy, including topics such as coal, petroleum, natural gas, oil shale, and electric power engineering
Frequency: Monthly, with annual cumulation
Availability: Subscription from National Technical Information Service
Companion Resource: DOE Energy Data Base

Title: Fusion Energy Update, 1979- .
Producer: U.S. Department of Energy
Scope: Abstracts and indexes literature on aspects of fusion energy, including plasma physics, magnetic fields and coils, cooling systems, radiation hazards, and power conversion systems
Frequency: Monthly, with annual cumulation

Availability: Subscription from National Technical Information Service
Companion Resource: DOE Energy Data Base

Title: Gastroenterology Abstracts and Citations, 1966-1978.
Producer: National Institute of Arthritis, Metabolism, and Digestive Diseases of the National Institutes of Health
Scope: Indexes and abstracts of world literature relevant to the nature, causes, and therapy of diseases of the gastrointestinal tract
Frequency: Monthly; ceased in 1978
Availability: Was free to National Institutes of Health grantees and contractors working in gastroenterology, and to medical school libraries and was available to depository libraries (HE20.3313: (v.nos.&nos.))

Title: Geothermal Energy Update, 1976- .
Producer: U.S. Department of Energy
Scope: Indexes and abstracts current literature related to geothermal energy
Frequency: Monthly, with annual cumulation
Availability: On subscription from National Technical Information Service (E1.66:(date))
Companion Resource: DOE Energy Data Base

Title: Government Reports Announcements and Index (*GRA&I*), 1946- .
Producer: National Technical Information Service, U.S. Department of Commerce
Scope: Indexes and abstracts government sponsored research and development, technical reports, engineering reports, and other analyses prepared by federal agencies, their contractors or grantees
Frequency: Semi-monthly, with annual cumulation
Availability: On subscription from the National Technical Information Service (C51.9/3:)
Companion Resources: NTIS Bibliographic Data Base. Also the following guides: *NTIS Subject Classification* (Past & Present) (PB-270 575); *A Reference Guide to the NTIS Bibliographic Data Base* (PR-253); *Environmental Microthesaurus* (PB-265 261); *Energy Microthesaurus* (PB-254 800); *Health Care Microthesaurus* (PB82-131715); *COSATI Subject Categories* (AD-612200); *COSATI Corporate Author Headings* (PB-198 275); *Computer Sciences Microthesaurus* (PB80-207814); *Social Sciences and Business Microthesaurus* (PB82-100066)

Title: HRIS [*Highway Research Information Service*] *Abstracts*, 1968- .
Producer: Transportation Research Board, National Research Council
Scope: Contains selected abstracts of literature on highway and urban transit research, prepared from computer tape records of the

Highway Research Information Service (HRIS) of the Transportation Research Board. Includes literature from the U.S. and other countries.
Frequency: Quarterly
Availability: On subscription from National Academy of Sciences, Transportation Research Board Publications Office, 2101 Constitution Avenue, NW, Washington, DC 20418
Companion Resource: HRIS Thesaurus Term List

Title: Habitat Preservation Abstracts, 1979- .
Producer: U.S. Fish and Wildlife Service
Scope: Announcement publication for information provided by the Habitat Preservation Programs of the U.S. Fish and Wildlife Service. These programs supply ecological information and technology related to environmental issues affecting fish and wildlife resources and their supporting ecosystems, including coal extraction and conversion, power plants, mineral development, water resource analysis, and information management.
Frequency: Irregular
Availability: U.S. Fish and Wildlife Service, One Gateway Center, Newton Corner, MA 02158; (617)965-5100, ext. 217; available to depository libraries (I49.18:H11/date)

Title: Highway Safety Literature, 1974- .
Producer: U.S. Department of Transportation, National Highway Traffic Safety Administration, Washington, DC 20590
Scope: Covers highway and traffic safety information
Frequency: Monthly
Availability: NHTSA, 400 Seventh Street, SW, Washington, DC 20590. Available to depository libraries (TD8.10:(nos.)).

Title: Hospital Literature Index, 1945- .
Producer: The American Hospital Association and the National Library of Medicine (NLM)
Scope: Literature on administration, financing and planning hospitals and related health care institutions; health manpower; and the administrative aspects of the medical, paramedical, and prepayment fields
Frequency: Quarterly, with annual cumulation
Availability: Subscription available from the American Hospital Association, Order Processing Dept., 840 North Lake Shore Drive, Chicago, IL 60611
Companion Resources: Hospital Literature Subject Headings; Hospital Literature Subject Headings Transition Guide to Medical Subject Headings,[6] and the Health Planning and Administration Data Base (available through the American Hospital Association and through NLM's MEDLINE)

Title: Index-Catalogue of Medical and Veterinary Zoology, 1953- .
Producer: Agricultural Research Service, U.S. Department of Agriculture

Scope: Provides in-depth coverage of the world's literature on human and animal parasitology. It covers parasitic protozoa, helminths, certain parasitic arthropods (ticks, mites, insects, copepods, isopods, linguatulids, and rhizocephalids), and miscellaneous parasites (mesozoans, molluscs, leeches, and turbellarians). The supplements are issued biennially and each has seven parts: 1. Authors: A-Z; 2. Protozoa; 3. Trematoda and Cestoda; 4. Nematoda and Acanthocephala; 5. Arthropoda and Miscellaneous Phyla; 6. Subject Headings and Treatment; and 7. Hosts. Entries are made in the various parts of the *Index-Catalogue* alphabetically by author, parasite, host, subject heading (e.g., Immunity, Monoclonal antibodies; Pheromones; etc.), and drug treatment. Information recorded includes: parasites and hosts reported, body location on or in the host, geographic distribution, parasite synonymy, taxonomic changes, classification, keys, drug treatment, and brief phrases describing subject matter content. Basic bibliographic information is included on each entry.
Frequency: Biennial
Availability: Recently discontinued by the federal government, the *Index-Catalogue* is now published by Oryx Press. Parts of the *Index-Catalogue* through Supplement 23 were available to depository libraries (A77.219/2:).

Title: Index Medicus, 1960- .
Producer: National Library of Medicine
Scope: Bibliography of current journal and selected monographic literature of biomedicine, worldwide
Frequency: Monthly; annual cumulation, *Cumulated Index Medicus*, is sold separately
Availability: On subscription from the Superintendent of Documents; available to depository libraries (HE20.3612:(v.nos.&nos.))
Companion Resources: MEDLARS (National Library of Medicine's Medical Literature Analysis and Retrieval System); *Medical Subject Headings (MeSH)*. Over 20 recurring bibliographies which are printed and distributed by nonprofit and professional organizations and governmental agencies. These bibliographies are produced by NLM from MEDLARS data base. Examples of the recurring bibliographies are *Anesthesiology Bibliography, Hospital Literature Index*, and *Psychopharmacology Bibliography*. Other companion publications include: *Bibliography of Medical Reviews, List of Journals Indexed in Index Medicus, NLM Current Catalog, Abridged Index Medicus*, and the like.

Title: International Nursing Index, 1966- .
Producer: American Journal of Nursing Company, in cooperation with the National Library of Medicine
Scope: Provides access to articles in over 200 nursing journals and articles pertinent to nursing published in 2,300 medical and allied health journals indexed by the National Library of Medicine

Frequency: Quarterly
Availability: On subscription from the American Journal of Nursing Company, 555 West Fifty-seventh Street, New York, NY 10019
Companion Resource: MEDLARS of the National Library of Medicine

Title: International Aerospace Abstracts
Producer: National Aeronautics and Space Administration and the American Institute of Aeronautics and Astronautics
Scope: Abstracts and indexes book, journal, conference literature, cover-to-cover journal translations, and selected foreign dissertations in fields related to aerospace research and technology. Complements *Scientific and Technical Aerospace Reports* (*STAR*), which focuses on aerospace report literature.
Frequency: Semi-monthly, with semi-annual cumulative indexes
Availability: On subscription from the American Institute of Aeronautics and Astronautics, Technical Information Service, 1290 Avenue of the Americas, New York, NY 10104
Companion Resources: NASA/RECON (NASA's data base); *NASA Thesaurus*

Title: Kidney Disease and Nephrology Index, 1975-1979.
Producer: National Institute of Arthritis, Metabolism, and Digestive Diseases
Scope: A bimonthly listing of literature on experimental and pathological nephrology, clinical urology, and including the Artificial Kidney Bibliography
Frequency: Bimonthly; ceased in 1979
Availability: Was available on subscription and to depository libraries (HE20.3318:(v.nos.&nos.))

Title: Limited Scientific and Technical Aerospace Reports (*LSTAR*), 1972/73- .
Producer: National Aeronautics and Space Administration
Scope: LSTAR, a companion abstract journal to *Scientific and Technical Aerospace Reports* (*STAR*), announces publicly available and unclassified aeronautics and space publications from over the world. *LSTAR* also announces NASA unclassified reports with availability limitations, as well as those that are security classified.
Frequency: Quarterly
Availability: Automatically distributed to NASA Center libraries. To receive *LSTAR*, other institutions must be registered with a contract with the NASA Scientific and Technical Information Facility and request receipt of *LSTAR* on form FF713A (green form).
Companion Resources: NASA/RECON (NASA's data base); *NASA Thesaurus*; *Confidential Scientific and Technical Aerospace Reports* (*CSTAR*)

Title: Monthly Catalog of United States Government Publications, 1895- .
Producer: Superintendent of Documents, U.S. Government Printing Office
Scope: The most comprehensive general index of federal government publications
Frequency: Monthly, with annual cumulated index
Availability: On subscription from the Superintendent of Documents; available to depository libraries (GP3.8:(date))
Companion Resources: GPO Monthly Catalog (*Monthly Catalog* online); *Library of Congress Subject Headings*

Title: NTIS Abstract Newsletters
Producer: National Technical Information Service, U.S. Department of Commerce
Scope: Individual weekly newsletters in 26 subject areas with summaries of recent, unclassified federally funded research
Frequency: Weekly, with index in last issue of the year
Availability: On subscription from NTIS; (C51.9/(nos.))
Companion Resource: NTIS Bibliographic Data File

Title: Neurosurgical Biblio-index, 1969- .
Producer: The American Association of Neurological Surgeons, in cooperation with the National Library of Medicine
Scope: A supplement to the *Journal of Neurosurgery*, produced through use of MEDLARS. Lists recent neurosurgically oriented articles from journals worldwide.
Frequency: Quarterly
Availability: Journal of Neurosurgery, Portsmouth Medical School, Hanover, NH 03755. Included as part of a subscription to Journal of Neurosurgery and cannot be purchased separately.

Title: Official Gazette of the United States Patent and Trademark Office, 1871- .
Producer: Commissioner of Patents and Trademarks, Washington, DC 20231
Scope: Contains abstracts of patents granted in the General and Mechanical, Chemical, and Electrical areas. Also includes reissue, design and plant patents.
Frequency: Weekly
Availability: Superintendent of Documents. Available to depository libraries (C21.5:(v./no.))

Title: Oncology Overviews
Producer: International Cancer Research Data Bank (ICRDB) Program of the National Cancer Institute
Scope: Oncology Overviews contain abstracts of recent articles dealing with topics in three major series: diagnosis and therapy; carcinogenesis; and cancer virology, immunology, and biology.
Frequency: Approximately 10 per year in each category

Availability: Available individually or as series subscriptions from the National Technical Information Service. A list of individual titles may be requested from NTIS.
Companion Resource: CANCERLINE, the online cancer information system of the ICRDB program

Title: Pesticides Abstracts, 1974- .
Producer: Office of Pesticides and Toxic Substances, Environmental Protection Agency
Scope: Abstracts of major worldwide literature on the effects of pesticides
Frequency: Monthly, with annual index
Availability: On subscription from the Superintendent of Documents; available to depository libraries (EP5.9:(v.nos.&nos.))

Title: Physical Fitness/Sports Medicine, 1978- .
Producer: The President's Council on Physical Fitness and Sports
Scope: A bibliography of citations selected from over 3,000 periodicals, and papers delivered at selected congresses. The citations have been retrieved from the MEDLARS data base of the National Library of Medicine.
Frequency: Quarterly
Availability: On subscription from the Superintendent of Documents; available to depository libraries (HE20.111:(v.nos.&nos.))

Title: Psychopharmacology Abstracts, 1961- .
Producer: U.S. Department of Health and Human Services, National Institute of Mental Health
Scope: New developments and research into the nature and causes of mental disorders and methods of treatment and prevention
Frequency: Quarterly, with annual cumulation
Availability: By subscription from the Superintendent of Documents; available to depository libraries (HE20.8109/2:(v.nos.&nos.))

Title: Publications Catalog of the U.S. Department of Health and Human Services
Producer: Office of the Secretary, Assistant Secretary for Public Affairs, U.S. Department of Health and Human Services, Washington, DC 20201
Scope: Lists the publications issued by various units of HHS during the preceding year. Publications listed are available from the Government Printing Office and the National Technical Information Service.
Frequency: Annually
Availability: Superintendent of Documents

Title: Publications of the National Bureau of Standards,[7] 1975- .
Producer: Technical Information and Publications Division, National Bureau of Standards

Scope: Provides abstracts of NBS publications in the bureau's periodicals, nonperiodical series, and contributions by NBS authors to non-NBS media during the past year
Frequency: Annually
Availability: For sale from the Superintendent of Documents. Available to depository libraries (C13.10:305/supp.no.).

Title: Scientific and Technical Aerospace Reports (*STAR*), 1963- .
Producer: National Aeronautics and Space Administration
Scope: All aspects of aeronautics and space research and development, supporting basic and applied research, and applications; aerospace aspects of earth resources, energy development, conservation, oceanography, environmental protection, and urban transportation. Abstracts and indexes technical reports of NASA, NASA contractors and grantees, federal government agencies, universities, private firms, and U.S. and foreign institutions; translations in report form; NASA-owned patents and patent applications; and dissertations and theses.
Frequency: Semi-monthly, with semi-annual cumulative indexes
Availability: On subscription from the Superintendent of Documents; available to depository libraries (NAS1.9/4:(v.nos.&nos.))
Companion Resources: NASA/RECON (NASA's data base); *NASA Thesaurus.*

Title: Selected Water Resources Abstracts, 1968- .
Producer: Office of Water Research and Technology, U.S. Department of the Interior
Scope: Abstracts of literature on water resources as treated in the life, physical, and social sciences; related engineering and legal aspects of the characteristics, supply condition, conservation, control use, and management of water resources
Frequency: Semi-monthly through 1982; monthly since Jan., 1983
Availability: On subscription from the National Technical Information Service; available to federal agencies, their contractors and grantees involved with water resources research upon request; available to depository libraries (I1.94/2:(date))
Companion Resource: Selected Water Resources Abstracts (SWRA) Data Base; *Water Resources Thesaurus*[8]

Title: Sexually Transmitted Diseases: Abstracts and Bibliography, 1979- .
Producer: Public Health Service, Department of Health and Human Services and Centers for Disease Control
Scope: Abstracts of journal articles chosen from MEDLINE, related to sexually transmitted diseases. Formerly titled *Abstracts for Current Literature on Venereal Disease.*
Frequency: Irregular
Availability: STD: Abstracts and Bibliography, Technical Information Services, Center for Prevention Services, Center for Disease Control, Atlanta, GA 30333. Also available to depository libraries (HE20.7002:Se9)

Title: Solar Energy Update, 1976- .
Producer: U.S. Department of Energy
Scope: Indexes and abstracts current literature, and information on research, development, and demonstration related to solar energy
Frequency: Monthly, with annual cumulation
Availability: On subscription from National Technical Information Service
Companion Resource: DOE Energy Data Base

Title: Sport Fishery Abstracts, 1955- .
Producer: Fish and Wildlife Service, U.S. Department of the Interior
Scope: Abstracts of current literature in sport fishery research and management
Frequency: Quarterly
Availability: On subscription from the Superintendent of Documents; available to depository libraries (I49.40/2:(v.nos.&nos.))

Title: Technical Abstract Bulletin (*TAB*)
Producer: Defense Technical Information Center (DTIC)
Scope: Indexes research, development, test and evaluation documents acquired by DTIC. Includes new classified and unclassified/limited scientific and technical reports received by DTIC.
Frequency: Bi-weekly, with annual index cumulations
Availability: Free to authorized DTIC users
Companion Resources: DTIC data bases

Urban Mass Transportation Abstracts, 1980- .
Producer: U.S. Department of Transportation, Urban Mass Transportation Administration, Transportation Research Information Center
Scope: Abstracts of research, development, and demonstrations sponsored and funded by the Urban Mass Transportation Administration
Frequency: Bi-monthly, with annual cumulation
Availability: Free from U.S. Department of Transportation, Urban Mass Transportation Administration, Transportation Research Information Center, Room 6428, 400 Seventh Street, SW, Washington, DC 20590; (202)426-9256. Available to depository libraries (TD7.10/2:(date)).

Title: Wildlife Review, 1935- .
Producer: Fish and Wildlife Service, U.S. Department of the Interior
Scope: Indexes and abstracts wildlife literature issued by the Fish and Wildlife Service. Includes topics such as conservation, plants, and wildlife.
Frequency: Quarterly
Availability: On subscription from the Superintendent of Documents; available to depository libraries (I49.17:(nos.))

REFERENCES

1. John B. Blake, ed., *Centenary of Index Medicus: 1879-1979* (Bethesda, MD: U.S. Department of Health and Human Services, Public Health Service, National Institutes of Health, National Library of Medicine, 1980) (NIH pub. no. 80-2068). Also: Burton W. Adkinson, *Two Centuries of Federal Information* (Stroudsburg, PA: Dowden, Hutchinson & Ross, 1978).

2. Adkinson, op. cit., pp. 121-25.

3. Ibid., pp. 124-25.

4. Ibid., pp. 123, 152.

5. U.S. Army, Research Institute for the Behavioral and Social Sciences, *Abstracts of ARI Research Publications, FY 1977* (Technical rept. 456) (Alexandria, VA: U.S. Army Research Institute for the Behavioral and Social Sciences, 1980).

6. Alice Dunlap, *Hospital Literature Subject Headings* (Chicago: American Hospital Assn., 1977); *Hospital Literature Subject Headings Transition Guide to Medical Subject Headings* (Chicago: American Hospital Assn., 1978).

7. Betty L. Bunis and Rebecca J. Morehouse, eds., *Publication of the National Bureau of Standards 1980 Catalog: Compilation of Abstracts and Key Word and Author Indexes* (NBS Special Publication 305 Supplement 12) (Washington, DC: U.S. Government Printing Office, 1981).

8. U.S. Department of the Interior, Office of Water Research and Technology, *Water Resources Thesaurus: A Vocabulary for Indexing and Retrieving the Literature of Water Resources Research and Development*, 3d ed. (OWRT IT-80/1) (Washington, DC: GPO, 1980) (I1.2:W29/2/980).

11

DATA BASES

Data bases share a common purpose with print resources such as books, journals, reports, indexes, and abstracts. Each has been created to facilitate the communication of information. Differing formats offer different modes of information storage and retrieval, and each format has distinct advantages and disadvantages. The advantages of machine-readable data base services are so considerable that the scientist or researcher who overlooks them may be seriously limiting his access to complete, current, and speedy information retrieval.

The advantages of utilizing information in machine-readable form include the fact that it offers fast retrospective and current searching; output in a number of formats, depending on user preference; full or partial bibliographic information, again depending on individual preference; manipulation of numeric data; and provision of searches tailored to specific information requests. Some disadvantages include the cost, which can be high with some data base services, and the need to construct a precise search statement to avoid retrieval of unwanted items or missing desired ones.

One frustration related to government machine-readable data bases results from their sheer numbers and decentralized production—that is, the difficulty of identifying an appropriate data base when one is needed. This chapter offers an introduction to available data base services, describes selected data bases and data files, and provides references to sources of further information.[1]

DATA BASE PRODUCTION

Data base production and use involves three basic elements: producers, distributors, and users.[2] The federal government and its agencies produce or sponsor a significant proportion of these data bases, and thus seek to make many of them available to the public. Other producers of machine-readable data bases include commercial organizations, technical and professional societies, and research institutes.

A second element is the data base vendor or distributor. These are the organizations that make data bases available to users via online systems, frequently packaging many data bases together. Data base distributors, like data base producers, may be either commercial or not-for-profit institutions. The three best known commercial data base vendors are System Development Corporation, DIALOG Information Services, Inc., and Bibliographic Retrieval Services, Inc. Each of these vendors leases a "package" of numerous data bases to its subscribers via their online systems known respectively as ORBIT, DIALOG, and BRS.[3] Government-produced data bases are included in each of these three vendors' package of data bases.

In many cases, a government agency acts as both data base producer and distributor of a data base produced in the course of its mission, or may offer alternative access to the data base by leasing it to commercial vendors, or to other government agencies. The NTIS Bibliographic Data Base is a good example of this practice. NTIS offers its own custom online search services conducted by NTIS staff, and also leases its data base to other organizations. Through leasing arrangements, the NTIS data base is availble from all three commercial vendors as well as from other government agencies, such as NOAA, which searches the NTIS data base along with several others as part of its own search services.

The third element of the data base industry is made up of data base users. These include both individuals and organizations. Most organizational data base users such as large academic and research libraries, corporations, and government agencies, subscribe to data base services and then make them available to their own clientele—scientists, researchers, scholars, etc.

BIBLIOGRAPHIC AND NUMERIC DATA BASES

Each of these two types of data bases offers a valuable information resource to the scientist and researcher. Both are collections of information in machine-readable form, but differences in the type of machine-readable information stored in them result in distinctions in their uses. A bibliographic data base is one that contains references to documents, and is used for literature searching. A computer search of a bibliographic data base will result in citations to documents, with the capability of providing full bibliographic information, as well as abstracts when they are available.

A numeric data base, on the other hand, is a machine-readable collection of data that are primarily numeric in nature. Examples of such numeric data include measures of physical or chemical properties, time, income, or prices. Sometimes called factual data bases, or "electronic handbooks," they are compilations of data, often compiled from the original literature or research results. In addition to providing search capabilities for locating and retrieving numeric data, these data base systems usually offer computational routines for analyzing and manipulating these data.

SCOPE OF DATA BASE SERVICES

Williams calculated that in 1979 there were 528 data bases* available in the United States, containing about one hundred forty-eight million records.[4] Although privately sponsored data bases outnumbered those generated by state and federal governments (about 25 percent were publicly supported in 1979), government money contributed significantly to the support of many non-government data bases.[5]

The number of available data bases jumped 75 percent between 1975 and 1979, accompanied by a surge in their use. The number of online searches actually quadrupled during the same 4-year period.[6] And, as data bases proliferate, keeping track of them becomes more difficult. No comprehensive source exists which lists all data files produced by the federal government. Their production and distribution is decentralized, linked to the missions and activities of numerous agencies.

Frequently created for internal agency use and analysis, some data files remain inaccessible to the public because the producing agency cannot afford to prepare them for public use. Nevertheless, the U.S. government has made an effort to organize broad categories of mission-oriented data files that are in the public domain and provide access to them. Their use involves a two-step process: 1) identifying the appropriate data base and 2) gaining access to the data base, either directly through its government producer, or indirectly through another distributor such as a vendor.

The remainder of this chapter is devoted to describing selected U.S. government data bases. Descriptions include information on their subject coverage, scope, and accessibility. In addition, tables 11.1 (pages 168-91) and 11.2 (pages 192-205) provide a brief listing of selected government data bases, both bibliographic and numeric. To identify additional data bases or to gather supplemental information on any of the data bases described, consult the guides listed in the bibliography at the end of this chapter.

(Text continues on page 206.)

---

*1982 edition of *Computer-Readable Databases: A Directory and Data Sourcebook*, which was edited by Williams in 1982, lists 773 bibliographic data bases.

TABLE 11.1
SELECTED GOVERNMENT BIBLIOGRAPHIC DATA BASES

DATA BASE NAME	SUBJECT
Agricola	Agriculture
Aquaculture	Literature pertinent to freshwater, brackish, or marine animals and plants
Air Pollution Technical Information Center	Literature on air quality and air pollution

TABLE 11.1 (cont'd)

PRODUCER	VENDOR*
U.S. Dept. of Agriculture	B,L,S
National Oceanic and Atmospheric Administration	L
EPA	

*B=BRS,L=Lockheed, S=SDC

(Table 11.1 continues on page 170.)

NOTE: L = Lockheed refers to DIALOG Information Services, Inc.

TABLE 11.1 (cont'd)

DATA BASE NAME	SUBJECT
AVLINE	Audiovisual instructional materials in health sciences
Bioethicsline	References to literature on bioethical topics such as euthanasia, human experimentation, and abortion
BIRS (Biological Information Retrieval System)	Biology

TABLE 11.1 (cont'd)

PRODUCER	VENDOR*
NLM	
NLM	
National Oceanic and Atmospheric Administration	

*B=BRS, L=Lockheed, S=SDC

(Table 11.1 continues on page 172.)

TABLE 11.1 (cont'd)

DATA BASE NAME	SUBJECT
CANCERLIT	Cancer
COLD REGIONS data base	Literature related to Antarctica, the Antarctic Ocean, subantarctic islands; snow, ice, and frozen ground; navigation on ice; civil engineering in cold regions; and behavior/operation of equipment in cold temperatures
DOE Energy Data Base (EDB)	Energy

TABLE 11.1 (cont'd)

PRODUCER	VENDOR*
NLM (distributor)	
Cold Regions Research and Engineering Laboratory, U.S. Army Corps of Engineers	S
U.S. Dept. of Energy	B,L,S

*B=BRS, L=Lockheed, S=SDC

(Table 11.1 continues on page 174.)

TABLE 11.1 (cont'd)

DATA BASE NAME	SUBJECT
DOE/STOR Bibliography for Flywheel Energy Systems	Flywheels, other electric hybrid vehicles
DOE/STOR Bibliography for molten Salts - General	Material properties of molten salts
EPA Reports System	Reports produced by EPA

TABLE 11.1 (cont'd)

PRODUCER	VENDOR*
U. S. Dept. of Energy	
U. S. Dept. of Energy	
NTIS	

*B=BRS, L=Lockheed, S=SDC

(Table 11.1 continues on page 176.)

TABLE 11.1 (cont'd)

DATA BASE NAME	SUBJECT
Epilepsyline	References to articles on epilepsy
ERDA/RECON	World-wide research and development reports related to energy
Fish and Wildlife Reference Service	American fish and wildlife

TABLE 11.1 (cont'd)

PRODUCER	VENDOR*
NLM	
U.S. Energy Research and Development Administration	
U. S. Fish and Wildlife Service	

*B=BRS, L=Lockheed, S=SDC

(Table 11.1 continues on page 178.)

TABLE 11.1 (cont'd)

DATA BASE NAME	SUBJECT
GPO	Monthly Catalog on-line: publications of Federal agencies, and U.S. Congress
Health Planning and Administration	References to literature on health planning, organization, financing, management and manpower
Highway Safety Literature	Traffic and motor vehicle safety

TABLE 11.1 (cont'd)

PRODUCER	VENDOR*
U. S. Government Printing Office	B,L,S
NLM	B
National Highway Traffic Safety Administration	

*B=BRS,L=Lockheed, S=SDC

(Table 11.1 continues on page 180.)

TABLE 11.1 (cont'd)

DATA BASE NAME	SUBJECT
HISTLINE (History of medicine Online)	References to literature on the history of medicine & related sciences
MEDLINE (Medlars online)	References to biomedical journal articles worldwide
Meteorological and Geoastrophysical Abstracts	Meteorological and geoastrophysical literature from both U.S. and foreign sources

TABLE 11.1 (cont'd)

PRODUCER	VENDOR*
NLM	
NLM	B,L
American Meteorological Society and National Oceanic and Atmospheric Administration	

*B=BRS,L=Lockheed, S=SDC

(Table 11.1 continues on page 182.)

TABLE 11.1 (cont'd)

DATA BASE NAME	SUBJECT
NASA STI	Worldwide technical reports on aerospace
National Clearinghouse for Mental Health Information Data Base	Mental health literature
National Index of Computer-readable Energy and Environmentally related Databases	Covers 1965-75 and identifies 4,103 federal databases

TABLE 11.1 (cont'd)

PRODUCER	VENDOR*
NASA	
National Clearinghouse for Mental Health Information	B
ERDA	

*B=BRS, L=Lockheed, S=SDC

(Table 11.1 continues on page 184.)

TABLE 11.1 (cont'd)

DATA BASE NAME	SUBJECT
Noise Information System	Literature related to all aspects of noise
NTIS Bibliographic Data File	Unclassified technical reports
Nuclear Safety Information Center	Nuclear power plants and research reactors

TABLE 11.1 (cont'd)

PRODUCER	VENDOR*
EPA	
NTIS	B,S,L
Oak Ridge National Labatory	

*B=BRS, L=Lockheed, S=SDC

(Table 11.1 continues on page 186.)

TABLE 11.1 (cont'd)

DATA BASE NAME	SUBJECT
Nuclear Structure References	Nuclear physics
OASIS (Atmospheric Information System)	Technical literature related to atmospheric, oceanic, and solid earth disciplines
PIE (Pacific Islands Ecosystems data base)	Literature on biological, ecological, physical, and socioeconomic aspects of the Pacific Islands

TABLE 11.1 (cont'd)

PRODUCER	VENDOR*
U. S. Dept. of Energy	
NOAA	
Office of Biological Services, Fish and Wildlife Service	S

*B=BRS, L=Lockheed, S=SDC

TABLE 11.1 (cont'd)

DATA BASE NAME	SUBJECT
Popline (Population Information Online)	References to literature on population, including reproductive biology, contraceptive technology, family planning, and demography
STAR	Aerospace technical report literature
Sea Grant Bibliographic Data Base	Access to All Sea Grant publications

TABLE 11.1 (cont'd)

PRODUCER	VENDOR*
NLM	
NASA	
Nat'l Sea Grant Depository, University of Rhode Island	

*B=BRS, L=Lockheed, S=SDC

(Table 11.1 continues on page 190.)

TABLE 11.1 (cont'd)

DATA BASE NAME	SUBJECT
Selected Water Resources Abstracts	Literature on all aspects of water resources
SERLINE	Serials and numbered congresses received by NLM
Solid waste information retrieval system	International literature on solid waste management
TRIS-On-Line (Transportation Research Services)	Literature and research in progress on transportation, highways, and related topics

TABLE 11.1 (cont'd)

PRODUCER	VENDOR*
U. S. Department of the Interior	
NLM	
EPA	
Transportation Research Board	L

*B=BRS, L=Lockheed, S=SDC

TABLE 11.2
SELECTED FACT-ORIENTED DATA BASES

DATA BASE NAME	SUBJECT
Agriculture/Weather	Acreages, weather conditions, yields, prices and consumption of U. S. crops
Cancerproj (Cancer Research Projects)	Ongoing Cancer Research
Carcinogenesis Bioassay Data System	Information on carcinogenesis bioassays
CHEMLINE	Nomenclature and structural fragments for chemical compounds

TABLE 11.2 (cont'd)

PRODUCER	VENDOR*
U. S. Dept. of Agriculture	
NLM	
National Cancer Institute	
NLM, Toxicology Information Program	

*B=BRS, L=Lockheed, S=SDC

(Table 11.2 continues on page 194.)

NOTE: L = Lockheed refers to DIALOG Information Services, Inc.

TABLE 11.2 (cont'd)

DATA BASE NAME	SUBJECT
Crystal Data Tape	Crystallographic data on 18,000 compounds
Current Research Information System (CRIS)	Abstracts of current research projects in agriculture
DOE/STOR Bibliography for molten Salts - Eutectic Data	Melting points and compositions of molten salts eutectic mixtures
Drug Research and Development Biological Data Processing	Information on the anticancer activity of synthetic compounds, plant and animal materials, and fermentation products

TABLE 11.2 (cont'd)

PRODUCER	VENDOR*
National Bureau of Standards	
U. S. Dept. of Agriculture	L
U. S. Dept. of Energy	
National Cancer Institute	

*B=BRS, L=Lockheed, S=SDC

(Table 11.2 continues on page 196.)

TABLE 11.2 (cont'd)

DATA BASE NAME	SUBJECT
Drug Research and Development Chemical Data Processing System	All chemicals tested for antitumor activity by the Division of Cancer Treatment of the NCI
Environmental Data Index (ENDEX)	Interdisciplinary environmental data
In Vitro Information System	Information about mutagenesis assays
Ionization Potential Data Base	Heats of Formation of Gaseous Ions

TABLE 11.2 (cont'd)

PRODUCER	VENDOR*
National Cancer Institute	
NOAA	
National Cancer Institute	
National Bureau of Standards	

*B=BRS, L=Lockheed, S=SDC

TABLE 11.2 (cont'd)

DATA BASE NAME	SUBJECT
Laboratory Animal Data Bank	Comparative data on laboratory animals used as experimental controls
National Drug Code Directory	Product trade names, products by National Drug Code, and other information on human prescription drugs and selected over-the-counter drugs
National Referral Center data base	A computerized directory of organizations offering specialized knowledge in science and social science

TABLE 11.2 (cont'd)

PRODUCER	VENDOR*
NLM	
H.H.S. (available from NTIS)	
NRC, Library of Congress	

*B=BRS, L=Lockheed, S=SDC

(Table 11.2 continues on page 200.)

TABLE 11.2 (cont'd)

DATA BASE NAME	SUBJECT
Oil and Hazardous Materials Technical Assistance Data Program (OHM - TADS Bibliographic Data Base)	Spill response efforts for oil or hazardous materials
Pesticides Active Ingredients	Registered active ingredients of selected pesticides
Pesticides Analytical Reference Standards	Standards for testing pesticides
Psychotropic Drugs	Information on the psychotropic effects of some 2,000 compounds

TABLE 11.2 (cont'd)

PRODUCER	VENDOR*
EPA	
Environmental Protection Agency	
Environmental Protection Agency	
National Institute of Mental Health	

*B=BRS, L=Lockheed, S=SDC

(Table 11.2 continues on page 202.)

TABLE 11.2 (cont'd)

DATA BASE NAME	SUBJECT
RANN pollutant file	File of hazardous chemicals
Smithsonian Science Information Exchange (SSIE)	Research in progress in all areas of life, physical, social, behavioral, and engineering sciences
STORET	Storage & Retrieval for Water Quality Data of Waterways, USA
Toxicology Data Bank	Chemical, physical, pharmacological and toxicological facts and data

TABLE 11.2 (cont'd)

PRODUCER	VENDOR*
National Science Foundation	
SSIE	B,L,S
Environmental Protection Agency	
NLM	

*B=BRS, L=Lockheed, S=SDC

(Table 11.2 continues on page 204.)

TABLE 11.2 (cont'd)

DATA BASE NAME	SUBJECT
TOXLINE	Human and animal toxicity studies, adverse drug reactions, environmental chemicals and pollutants
TSCA (Toxic Substances Control Act)	A chemical dictionary file covering all chemicals in the TSCA Initial Inventory of 1976 and 1st supplement
World Fertilizer Market Information Service	Production, consumption & trade by country and product

TABLE 11.2 (cont'd)

PRODUCER	VENDOR*
NLM	
NTIS	
Tennesee Valley Authority	

*BRS, L=Lockheed, S=SDC

NATIONAL TECHNICAL INFORMATION SERVICE (NTIS)

The National Technical Information Service (NTIS) is the central source for information storage, retrieval, and sale of U.S. government-sponsored research, development, and engineering reports. NTIS also handles foreign technical reports, and other reports of national or local government agencies, their contractors, or grantees.

The National Technical Information Service collection contains more than one million titles in scientific as well as numerous other subject areas. Because of the diverse activities of federal agencies, the collection includes topics not just in the sciences, but in all areas of endeavor. The NTIS collection of unclassified/unlimited scientific and technical reports is bibliographically accessible through a machine-readable data base, and also through print sources produced from this data base. In most cases, either hard copy or microfiche copy of cited reports may be purchased from NTIS.

NTIS is also a central source for federally generated computer data files. In an effort to expand public access to machine-readable data files of federal agencies, NTIS has pursued the following activities:

Cataloging and indexing government machine-readable statistical data files.

Providing online access to federally produced data bases.

Expanding the NTIS collection of computerized data files and software.

Providing reference and tabulation services related to federal statistical data files.

NTIS seeks to provide online access to federal data bases that are not accessible from commercial vendors. Their current collection includes more than one thousand two hundred data files and computer programs in diverse subjects from more than one hundred federal agencies, available from NTIS for sale or lease. NTIS periodically publishes a *Directory of Computer Software and Related Technical Reports*, a guide to federally produced machine-readable software and related technical reports which are available to the public. NTIS also offers statisfical reference and tabulation services through their Statistical Data Reference Service and Statistical Data Tabulation Service.

The NTIS Bibliographic Data Base offers all of the research summaries and bibliographic citations announced by NTIS in its role as clearinghouse for government-sponsored research and technical reports. As such, it covers numerous disciplines and subject areas. Three agencies contribute the bulk of the documents listed in the data base: the U.S. Department of Energy, U.S. Department of Defense, and the National Aeronautics and Space Administration.

Because the NTIS Data Base represents documents from a number of agencies, several thesauri are used to determine the controlled vocabulary terms used for indexing reports. A list of these is given in an NTIS brochure, *A Reference Guide to the NTIS Bibliographic Data Base.*[7]

The NTIS Bibliographic Data Base is accessible through commercial systems (BRS, DIALOG, and ORBIT) or through lease from NTIS. Other data bases available for lease from NTIS include Energy Data Base, Patent Data File, Selected Water Resources Abstracts Data Base, AGRICOLA Data Base, Aquatic Sciences and Fisheries Abstracts Data Base, and the National Science Foundation Summaries of Projects completed. Each of these is available for lease from NTIS and can also be searched directly through the NTIS online search services (NTISearch).

Another service that draws from the NTIS data base is the NTIS series, Published Searches, packaged searches on selected topics of wide interest. More than twenty two hundred of these specially priced searches are available, each consisting of bibliographic citations and abstracts. NTIS publishes a master list of Published Searches as well as a series of minicatalogs.

DEFENSE TECHNICAL INFORMATION CENTER (DTIC)

The mission of the Defense Technical Information Center (DTIC) is to disseminate scientific and technical information for Department of Defense (DoD) personnel and contractors, as well as for other government agency personnel and contractors.[8] The DTIC acts as a central repository within the Department of Defense for interchange of scientific and technical research information.

The DTIC computerized information system entails four distinct data bases:

1) The Research and Development Program Planning Data Base, listing projects which forecast and propose future research activities.
2) The Research and Technology Work Unit Information System, which documents research projects currently being performed by the DoD or its contractors.
3) The Technical Reports Data Base, citing completed research projects.
4) The Independent Research and Development Data Base, listing research in progress jointly sponsored by DoD and industry.

The topics covered in these data bases include aeronautics, missile technology, space technology, navigation, and nuclear science. Less obviously associated with national defense, but also represented are biology, chemistry, energy, environmental sciences, oceanography, computer sciences, sociology, and human factors engineering.

DTIC ON-LINE

The Defense RDT&E[9] On-Line System (DROLS) is a network of remote terminals, allowing online access to the four DTIC data bases. DROLS is available only through DTIC. Any registered DTIC users are eligible for terminal hookup to DROLS and pay only for hardware and communication line costs.

Because of the type of information in which DTIC specializes, users must qualify and register for access to restricted materials. Organizations such as universities, libraries, and government agencies may register for service provided they are qualified in terms of defense contract work. Access to unclassified/ unlimited technical reports is available to the general public through the NTIS *Government Reports Announcements and Index* (print), or online from the NTIS Bibliographic Data File. DTIC registered users are eligible for several free services, including literature searches of DTIC's four data bases, biweekly selective dissemination of information, and subject searches performed online by DTIC staff.

U.S. DEPARTMENT OF ENERGY (DOE)

The Federal Energy Data Index (FEDEX) is a machine-readable bibliographic file containing indexes and abstracts for all Energy Information Administration (EIA) publications. In addition, FEDEX separately abstracts the indexes individual tables and graphs in each EIA publication. The subject content of EIA publications, tables, and graphs includes statistical data on energy production, consumption, prices, availability of resources, and supply-and-demand projections. Two publications are produced via the FEDEX data file: *EIA Publications Directory: A User's Guide*, which contains abstracts of EIA publications, and *EIA Data Index: An Abstract Journal*, with abstracts of tables and graphs.[10]

FEDEX can be searched online through the Department of Energy RECON system and through BRS. Individual FEDEX searches also are available through the National Energy Information Center Affiliate of the University of New Mexico. Contact National Energy Information Center Affiliate, University of New Mexico, 2500 Central Avenue, SE, Albuquerque, NM 87131; (505)846-2383.

A second Department of Energy data base is the Energy Data Base (EDB), which focuses on world scientific and technical literature on energy. Compiled by the DOE Technical Information Center, this data base provides citations and abstracts for energy publications from the public and private sectors. Subjects covered include solar and geothermal energy, fuels, nuclear and fusion energy, electric power engineering, and many others. The EDB is available on magnetic tape from the National Technical Information Center and is accessible via BRS, ORBIT, and DIALOG as well as the Department of Energy's online system, DOE/RECON.

NATIONAL AERONAUTICS AND SPACE ADMINISTRATION (NASA)

The National Aeronautics and Space Administration's computerized data base contains citations to over two million documents related to aeronautics and space. NASA/RECON, an online interactive system, provides direct rapid information retrieval from the computerized NASA data base. A literature search service provides access not only to the NASA data base, but also to more than one hundred non-NASA scientific and technical data bases.

NASA SCIENTIFIC AND TECHNICAL INFORMATION DATA FILE

The NASA computerized data file provides bibliographic data identifying documents of government organizations, universities, research laboratories, corporations, and commercial publishers in the United States and in more than two hundred foreign countries. The types of documents cited in the data file include: NASA-owned patents and patent applications, journal articles, books, pamphlets, monographs, conference proceedings, papers from scientific meetings, translations, non-NASA reports, dissertations, and theses. Documents in the collection have been indexed according to the controlled vocabulary in the *NASA Thesaurus*, providing the basis for computerized current awareness and search services.

NASA/RECON

RECON (Remote Console) is an online, interactive computerized information search and retrieval system. Its purpose is to enable users at remote locations to directly access the NASA central data base. The major document series searchable on RECON include: *STAR, International Aerospace Abstracts, Limited Scientific and Technical Aerospace Reports*, NASA Contracts Data File, *Computer Program Abstracts, NASA Tech Briefs*, and the NASA Library Collection. The RECON network is made up of more than thirty terminals at fixed and mobile locations, and eligible organizations can apply for terminal service. For further information, consult "The NASA/Recon [sic] Search System," by Robert F. Jack (*Online* 6 [November 1982]: 40-54).

LITERATURE SEARCH SERVICE

Individuals without RECON access may request individual literature searches of the NASA data base, the Defense Documentation Center data bank, and a number of commercial data bases through NASA's Scientific and Technical Information Office. NASA literature searches are available to NASA contractors and government agencies, but not to the general public. Individuals in the latter group may search the NASA data base through the services of NTIS, the European Space Agency, and NASA-sponsored Industrial Applications Centers in six U.S. locations.

NATIONAL AGRICULTURAL LIBRARY

The National Agricultural Library data base, AGRICOLA (Agricultural On-Line Access), is a family of data files with indexes to world journal and monographic literature, and U.S. technical reports on agriculture, agricultural economics, food and nutrition, and related topics. AGRICOLA, formerly called CAIN, consists of a number of subfiles, including indexed material from the American Agricultural Economics Documentation Center, the Food and Nutrition Education and Information Materials Center, the U.S. Department of

Agriculture Brucellosis file, Abstracts of Environmental Impact Statements, the Energy and Agriculture subfile, 4-H subfile, and Extension publications subfile.

AGRICOLA is available commercially from DIALOG, SDC, and BRS. U.S. Department of Agriculture personnel can access the data base through the department's Technical Information Systems Unit Reference Branch, Library Operations Division, Technical Information Systems, Science and Education Administration, U.S. Department of Agriculture, Beltsville, MD 20705; (301)344-3755. In addition, the National Technical Information Service sells AGRICOLA magnetic tapes for an annual fee per tape. Purchasers receive twelve monthly updates.

NATIONAL LIBRARY OF MEDICINE (NLM)

The National Library of Medicine is considered as one of three national libraries in the United States (along with the Library of Congress and the National Agricultural Library), and is the world's largest single-discipline research library. It serves as a source of medical information for individuals and organizations engaged in medical education, research, and services. The NLM maintains exhaustive collections in more than forty biomedical subject areas, and significant collections in related areas such as chemistry, physics, zoology, botany, psychology, and instrumentation. Its holdings include over 1.5 million books, journals, technical reports, documents, theses, pamphlets, microfilms, pictorial, and audiovisual materials.

The NLM offers extensive computerized literature retrieval services, the best known of which are probably MEDLARS and MEDLINE. MEDLARS (Medical Literature Analysis and Retrieval System) contains citations to health science related books and journal articles published since 1965. Most of these references have also been published via MEDLARS in printed NLM indexes and bibliographies such as *Index Medicus*. MEDLINE (MEDLARS online) contains citations (and many abstracts) for biomedical journal literature from the current and two previous years, and for chapters and articles from selected monographs. Retrospective files allow searching back to 1966. Coverage is not only of the United States, but also seventy foreign countries.

The NLM online search services are coordinated through seven Regional Medical Libraries, each responsible for a geographic region of the U.S., listed below. The NLM online network provides searches at more than thirteen hundred universities, medical schools, hospitals, government agencies, and commercial organizations across the country.

NLM Regional Medical Libraries

Region 1: Northeast Regional Medical Library Program (Connecticut, Delaware, Maine, Massachusetts, New Hampshire, New Jersey, New York, Pennsylvania, Rhode Island, Vermont, and Puerto Rico)
New York Academy of Medicine Library
Two East 103rd Street
New York, NY 10029

Region 2: Mid-Atlantic Regional Medical Library Program (Alabama, District of Columbia, Florida, Georgia, Maryland, Mississippi, North Carolina, South Carolina, Tennessee, Virginia, and West Virginia)
University of Maryland Health Sciences Library
111 South Green Street
Baltimore, MD 21201

Region 3: Midwest Regional Medical Library (Illinois, Indiana, Iowa, Kentucky, Michigan, Minnesota, North Dakota, Ohio, South Dakota, and Wisconsin)
Library of the Health Sciences
University of Illinois at the Medical Center
1750 West Polk Street
Chicago, IL 60612

Region 4: Midcontinental Regional Medical Library Program (Colorado, Kansas, Missouri, Nebraska, Utah, and Wyoming)
Library of Medicine
University of Nebraska Medical Center
Omaha, NE 68105

Region 5: South Central Regional Medical Library Program (Arkansas, Louisiana, New Mexico, Oklahoma, and Texas)
University of Texas Health Science Center
5323 Harry Hines Boulevard
Dallas, TX 75235

Region 6: Pacific Northwest Regional Health Sciences Library (Alaska, Idaho, Montana, Oregon, and Washington)
University of Washington Health Sciences Library
Seattle, WA 98195

Region 7: Pacific Southwest Regional Medical Library Service (Arizona, California, Hawaii, and Nevada)
Biomedical Library
Center for the Health Sciences
University of California at Los Angeles
Los Angeles, CA 90024

MEDLARS and MEDLINE are not the only data bases available through NLM's online network. Users also have access to TOXLINE, CHEMLINE, Toxicology Data Bank, and the Registry of Toxic Effects of Chemical Substances, each of which is discussed in the following pages. Other data bases available through NLM's online network include:

AVLINE (Audiovisuals On-Line)—references to audiovisual teaching packages in the health sciences

Health Planning and Administration—references to literature on health planning, organization, financing, management, manpower and related topics.

HISTLINE (History of Medicine Online)–references to literature on the history of medicine and related sciences

CANCERLIT (Cancer Literature)–references to literature on cancer worldwide

CANCERPROJ (Cancer Research Projects)–summaries of cancer research in progress, covering the current and previous two years

CLINPROT (Clinical Cancer Protocols)–summaries of clinical investigations of new anticancer agents and treatments

BIOETHICSLINE–references to literature on topics such as euthanasia, human experimentation, abortion, and other bioethical topics

EPILEPSYLINE–references to literature on epilepsy

POPLINE (Population Information Online)–citations to literature related to population, such as reproductive biology, applied research in contraceptive technology, family planning, and demography[11]

The National Library of Medicine Toxicology Information Program (TIP) was established in 1967 to provide toxicology information services for the scientific community, and to develop machine-readable toxicology data banks. TIP's information services include those provided by the Toxicology Information Response Center. Their computerized data bases include TOXLINE, CHEMLINE, the *Registry of Toxic Effects of Chemical Substances*, the Toxicology Data Bank, and the Laboratory Animal Data Bank.

The Toxicology Information Response Center (TIRC) offers toxicology information and reference services related to individual chemicals, chemical classes, and a variety of toxicology-related topics. The center offers query response services, publications, and online information retrieval services. Literature searches tailored to specific user requests, and staff-prepared information packages on current topics are available on a fee-for-service basis. More than 100 data bases can be searched online at the center.

TOXLINE (TOXicology information onLINE) contains more than four hundred thousand references to human and animal toxicity studies, effects of environmental chemicals and pollutants, adverse drug reactions, and analytical methodology. TOXLINE also contains information on ongoing toxicology and epidemiology research projects from the Smithsonian Science Information Exchange Data Base, and offers selective dissemination of information services.

CHEMLINE (CHEMical dictionary onLINE) is a chemical dictionary file capable of searching and retrieving chemical substance names representing over four hundred forty thousand unique substances. The CHEMLINE data file contains Chemical Abstracts Service Registry Numbers, molecular formulas, preferred chemical index nomenclature, generic and trivial names derived from the Chemical Abstracts Service Nomenclature File, other NLM data files with information on a chemical substance being searched, and Ring Information.

The *Registry of Toxic Effects of Chemical Substances* (formerly called the *Toxic Substances List*) offers toxicity information, chemical identification, standards, and regulations on about forty-four thousand substances. The registry is prepared annually by the National Institute for Occupational Safety and Health, and updated quarterly.

The Toxicology Data Bank offers access to chemical, pharmacological, and toxicological information and data from more than 100 major textbooks and handbooks. Over sixty data elements are included in the data bank, including synonyms, chemical and physical properties, molecular formulas, and Chemical Abstracts Service Registry numbers.

The Laboratory Animal Data Bank provides access to information and baseline data on control animals related to hematology, clinical chemistry, pathology, environment and husbandry, and growth and development. About one million observations are recorded in the data bank, voluntarily contributed by government agencies, the pharmaceutical industry, and institutions throughout the United States.

TOXLINE, CHEMLINE, Toxicology Data Bank, and the Registry of Toxic Effects of Chemical Substances are accessible online via terminals connected to the National Library of Medicine central computer facility, and by telephone-based communication linkage across the country. About thirteen hundred academic, commercial, and government organizations offer access to these data bases. The Laboratory Animal Data Bank is available both offline and online. Offline users can request searches by telephone, and receive their searches by mail. Online access is available direct to the computer by telephone via TYMNET. Charges for use of each of these data bases vary.

NATIONAL OCEANIC AND ATMOSPHERIC ADMINISTRATION (NOAA)

The National Oceanic and Atmospheric Administration specializes in information about the atmosphere, earth, oceans, sun, and space. The Environmental Data and Information Service (EDIS) is the unit within NOAA that provides computerized information retrieval services. EDIS not only provides access to NOAA's Environmental Data Index data base, but also provides computer searches of other environmental data files, published literature, research in progress, SDI services, and printed products of literature searches.

The Environmental Data Index, called ENDEX, is a machine-readable, interdisciplinary file of environmental data. ENDEX offers three major components: descriptions of data collection efforts; detailed inventories of large, heavily used files; and descriptions of data files. Each of these gives descriptions of data from the EDIS national data centers as well as non-EDIS sources. An example of a data file searchable through ENDEX is the Environmental Data Base Directory, a computerized inventory of 12,000 environmental data bases in federal, state, and local governments, academic institutions, and private industry in the U.S. and Canada.

Each ENDEX data base and the range of EDIS services have been described in *Guide to NOAA's Computerized Information Retrieval Services.*[12] ENDEX data bases are searchable through EDIS and many of the NOAA libraries or information centers.

In addition to searches of ENDEX, EDIS provides searches of more than one hundred bibliographic data bases, both subject oriented and cross-disciplinary. The DOE/Energy Data Base, Management Contents, MEDLINE, the NTIS Bibliographic Data File, and Selected Water Resources Abstracts represent a sampling of the wide range of data bases relevant to atmospheric, environmental, and marine studies that can be searched through EDIS. Selective dissemination of information searches are also available for many data bases.

NOAA personnel receive EDIS searches free of charge. NOAA contractors and non-NOAA users are charged on a cost-recovery basis. Price lists are available from the User Services Branch, D822, NOAA Library and Information Services Division, 6009 Executive Blvd., Rockville, MD 20852; (301)443-8330. Searches may be requested in person, by telephone, or by sending a NOAA "Request for Computer Search Services" form. The *Guide to NOAA's Computerized Information Retrieval Services* lists numerous offices through which NOAA personnel may direct their requests.

PATENT AND TRADEMARK OFFICE

A division of the U.S. Patent and Trademark Office, the Office of Technology Assessment and Forecast (OTAF) has constructed a master data base of all U.S. patents. The data elements available in this machine-readable file include:

Patent subclasses

Subclasses related to the Standard Industrial Classification System in 55 Product Fields and Product Field combinations

Type of ownership at the time of issue and country or state of residence of the inventor (for patents issued since 1963)

Date of filing application (for patents issued since 1963)

Name of corporate owner (for patents issued since 1969)

Patent title (since 1969)

Inventor's name and address (since 1975)

Field of search and references cited in the Patent Office search (since 1975)

This data base is employed to prepare custom reports to satisfy specific data needs, with the option of various presentation formats such as charts, tables, and graphs.

OTAF offers several custom report programs (called PAT-TAF Custom Reports) which provide a wide variety of data for any given patent grouping. These standard format reports are less expensive than individually tailored searches. They include the "Technology Profile" report, the "Organizational Profile" report, and the "SIC Product Field" reports.

The "Technology Profile" report is a detailed profile for a specific technology or groupings of technology. It can include information on rate and origin of patenting, types of assignment and specific assignment, patent numbers and titles, or names and addresses of inventors. The "Organizational Profile" report focuses on patent activity across classes and subclasses for any organization, group of organizations, state, country, or grouping of states or countries. "SIC Product Field" reports provide listings of certain Standard Industrial Classification (SIC) Product Fields in conjunction with patent subclasses. All Custom Reports are charged on a cost reimbursable basis, for which estimates can be obtained before contracting the service.

U.S. GEOLOGICAL SURVEY

The U.S. Geological Survey has published a descriptive listing of machine-readable data bases maintained and used by that organization. Based on a comprehensive inventory of data bases, *Scientific and Technical, Spatial, and Bibliographic Data Bases of the U.S. Geological Survey, 1979* includes information on the subject coverage, availabiity, geographic coverage, and users of each data base.[13] Access to some of the data bases is open to any individual or organizational users outside the survey; others are limited to in-house use or to certain categories of users. In each case, a contact person is given. The listing includes indexes by data base acronyms, contact persons, geographic coverage, keywords, and data base names.

NATIONAL CENTER FOR HEALTH STATISTICS (NCHS)

The National Center for Health Statistics (NCHS) produces data on the health of Americans. The center's major data collection programs are described below:

Vital Registration Program: data on births, deaths, marriages, and divorces in the U.S.

National Survey of Family Growth: data based on a 5-stage probability sample of women 15-44 years old, with children at home

Health Interview Survey: a continuing national sample survey in which health data are collected through personal household interviews

Health Examination Survey: a continuing national sample survey in which health data are collected through physical examinations, tests, and measurements

Health and Nutrition Examination Survey: collection of data on the nutritional status of Americans through physical examinations, tests, and measurements

Master Facility Inventory: a comprehensive file of inpatient health facilities in the U.S.

Hospital Discharge Survey: a continuing national sample survey of short-stay hospitals in the U.S.

National Nursing Home Survey: data on nursing homes, their residents and staff

National Ambulatory Medical Care Survey: a continuing national probability sample of ambulatory medical encounters

Data from these surveys and studies are presented in a variety of publications, including the *Vital and Health Statistics* series. Additional data collected by the center are available on public-use data tapes and unpublished tabulations. Two publications describing these products may be requested free of charge: *Catalog of Public Use Data Tapes from the National Center for Health Statistics* and *Data Systems of the National Center for Health Statistics.* Questions about published and unpublished center data should be sent to: Scientific and Technical Information Branch, Division of Data Services, NCHS, 3700 East-West Highway, Hyattsville, MD 20782; (301)436-NCHS.

ASSOCIATION OF PUBLIC DATA USERS

The Association of Public Data Users seeks to promote acquisition and use of public data on computer tape, as well as providing information about public data availability. Membership is open to institutions – academic, government, private and public. For information, write the association at 1601 North Kent Street, Suite 900, Arlington, VA 22209; (703)525-1480.

REFERENCES

1. A data base, as it is commonly used in the context of librarianship, refers to computer-readable bibliographic and/or non-bibliographic data which, usually, can be searched online. Data file usually implies, non-bibliographic machine-readable data such as census data. In any event, the terminology is imprecise and NTIS, for instance, tends to use data base and data file interchangeably. Both terms, however, imply computer-readable storage of data where data is stored on media such as magnetic tapes and magnetic disks.

2. Roger Christian, *The Electronic Library: Bibliographic Data Bases, 1978-79* (White Plains, NY: Knowledge Industry Publications, Inc., 1978), p. 5.

3. DIALOG Information Retrieval Services
   3460 Hillview Avenue
   Palo Alto, CA 94304

   ORBIT Information Retrieval System
   2500 Colorado Avenue
   Santa Monica, CA 90406

   Bibliographic Retrieval Services, Inc.
   1200 Route 7
   Latham, NY 12110

4. Martha E. Williams, "Database and Online Statistics for 1979," *ASIS Bulletin* 7 (December 1980): 27.

5. Ibid., pp. 28-29.

6. Ibid., p. 29.

7. U.S. Department of Commerce, National Technical Information Service, *A Reference Guide to the NTIS Bibliographic Data Base* (Springfield, VA, 1980) (NTIS-PR 253).

8. Formerly the Defense Documentation Center.

9. Research, Development, Test, and Evaluation.

10. U.S. Energy Information Administration, *EIA Publications Directory, A User's Guide* (Washington, DC: U.S. Government Printing Office, 1980- . Quarterly) (E3.27:(nos.)); U.S. Energy Information Administration, *EIA Data Index: An Abstract Journal* (Washington, DC: U.S. Government Printing Office, 1980- . Semiannual) (E3.2:D26); and Catherine Grissom, Shelly Prosser, and Eulalie Brown, "Federal Energy Data Index," *Proceedings of the 43rd ASIS Annual Meeting* 17 (1980): 200-3.

11. U.S. Department of Health and Human Services, Public Health Service, National Institutes of Health, *Medlars: The Computerized Literature Retrieval Services of the National Library of Medicine* (NIH publication no. 81-1286) (Washington, DC: U.S. Government Printing Office, 1981).

12. U.S. Department of Commerce, National Oceanic and Atmospheric Administration, Environmental Data and Information Service, *Guide to NOAA's Computerized Information Retrieval Services* (Washington, DC: U.S. Government Printing Office, 1979) (C55.208:C73).

13. U.S. Department of the Interior, Geological Survey, *Scientific and Technical, Spatial, and Bibliographic Data Bases of the U.S. Geological Survey, 1979* (Arlington, VA: Geological Survey, 1979) (I19.4/2:817).

BIBLIOGRAPHY

GENERAL GUIDES

Boyle, Jeanne E. "U.S. Government Produced Machine Readable Files." *Documents to the People* 9 (July 1981): 153-4.

A Directory of Computerized Data Files & Related Technical Reports. Springfield, VA: National Technical Information Service, 1980 (PB 80-217003).

Directory of Federal Statistical Data Files. Springfield, VA: National Technical Information Service, 1981 (PB 81-13175).

Directory of Online Information Resources. 9th ed. Rockville, MD: CSG Press, 1982.

Directory of Online Databases. Santa Monica, CA: Cuadra Associates, 1979- . Quarterly.

Schmittroth, John, Jr., ed. *Encyclopedia of Information Systems and Services.* 5th ed. Detroit: Gale Research Co., 1982.

Slamecka, Vladimir, and Davis B. McCarn. *The Information Resources and Services of the United States: An Introduction for Developing Countries.* Washington, DC: U.S. Government Printing Office, 1979 (S1.2:R31/3).

U.S. Executive Office of the President. Federal Coordinating Council for Science, Engineering and Technology. *Directory of Federal Technology Transfer.* 2d ed. Washington, DC: U.S. Government Printing Office, 1977.

Williams, Martha E. et al., comps. *Computer-Readable Databases: A Directory and Data Sourcebook.* Washington, DC: American Society for Information Science, 1982.

SPECIALIZED GUIDES

Argonne National Laboratory. Division of Environmental Impact Studies. *Survey of Biomedical and Environmental Data Bases, Models, and Integrated Computer Systems at Argonne National Laboratory.* Springfield, VA: National Technical Information Service, 1978 (E1.28:ANL/ES-65).

Heller, S. R., and G. W. A. Milne. *EPA/NIH Mass Spectral Data Base.* Washington, DC: U.S. Government Printing Office, 1978 (C13.48:63).

Huber, Ernest E. et al. *Inventory of Sources of Computerized Ecological Information.* Springfield, VA: National Technical Information Service, 1978 (E1.28:ORNL-5441).

"The Information Cycle." (¾-inch videotape cassette. Color. 20 min.). On loan from: Educational Resources Division, Training and Education Branch, U.S. Department of Agriculture, Technical Information Systems, Room 408, NAL Building, Beltsville, MD 20705; (301)344-3937. Describes AGRICOLA and CRIS data bases, and provides other information about National Agricultural Library services.

Miller, Charles W., and Rodney H. Strand. *Meteorological Data Bases Available for the United States Department of Energy Oak Ridge Reservation.* (Environmental Sciences Division Pub. No. 1184). Oak Ridge, TN: Oak Ridge National Laboratory, 1978 (E1.28:ORNL/TM-6358).

Nardone, John. *Computerized Material Property Data Information System.* Springfield, VA: NTIS, 1976 (D4.11/2:31).

U.S. Department of Agriculture. Science and Education Administration. *AGRICOLA User's Guide.* (Agricultural Reviews and Manuals: ARM-H-7). Beltsville, MD: U.S. Department of Agriculture, 1979 (A106.12:H-7).

U.S. Department of Commerce. National Technical Information Service. *A Reference Guide to the NTIS Bibliographic Data Base.* Springfield, VA: National Technical Information Service, 1980 (NTIS-PR 253).

U.S. Department of Commerce. National Oceanic and Atmospheric Administration. Environmental Data and Information Service. *Guide to NOAA's Computerized Information Retrieval Services.* Washington, DC: U.S. Government Printing Office, 1979 (C55.208:C73).

U.S. Department of Commerce. National Oceanic and Atmospheric Administration. Environmental Data Service. *User's Guide to ENDEX/OASIS: Environmental Data Index and the Oceanic and Atmospheric Scientific*

Information System. (Key to Oceanic and Atmospheric Information Sources No. 1). Washington, DC: U.S. Government Printing Office, 1976 (C55.219/5:1).

U.S. Department of Energy. Technical Information Center. *Energy Data Base: Subject Coverage, Literature Coverage, Data Elements, and Indexing Practices*. (DOE/TIC-4608). Oakridge, TN, 1981.

U.S. Department of Energy. Technical Information Center. *Energy Information Data Base: Subject Thesaurus*. (DOE/TIC-7000-R4). Springfield, VA: National Technical Information Service, 1979 (E1.55:979).

U.S. Department of Health, Education, and Welfare. National Cancer Institute. *Directory of Cancer Research Information Resources*. Springfield, VA: National Technical Information Service, 1979 (PB-293 187).

U.S. Department of Health, Education, and Welfare. Public Health Service. Office of Health Research, Statistics, and Technology. National Center for Health Statistics. *Data Systems of the National Center for Health Statistics*. (DHEW Pub. No. PHS 80-1247). Hyattsville, MD: The Center, 1980.

U.S. Department of Health and Human Services. Public Health Service. National Institutes of Health. *Medlars: The Computerized Literature Retrieval Services of the National Library of Medicine*. (NIH Pub. No. 81-1286). Washington, DC: U.S. Government Printing Office, 1981.

U.S. Department of the Interior. Office of Library and Information Services. *Information Sources and Services Directory*. Washington, DC: The Department, 1979 (I1.2:In3/2).

U.S. Department of the Interior. Geological Survey. *Scientific and Technical, Spatial, and Bibliographic Data Bases of the U.S. Geological Survey, 1979*. (Geological Survey Circular 817). Arlington, VA: Dept. of the Interior, Geological Survey, 1979 (I19.4/2:817).

U.S. Energy Information Administration. *EIA Data Index: An Abstract Journal*. Washington, DC: U.S. Government Printing Office, 1980- . Semiannual (E3.2:D26).

U.S. Energy Information Administration. *EIA Publications Directory, A User's Guide*. Washington, DC: U.S. Government Printing Office, 1980- . Quarterly (E3.27:(nos.)).

U.S. Federal Energy Administration. *Energy Information in the Federal Government: Energy Information Locator System*. Springfield, VA: National Technical Information Service, 1975 (PB-246 703).

U.S. General Accounting Office. *Federal Information Sources and Systems 1980: A Directory Issued by the Comptroller General*. (1980 Congressional Source Book Series). Washington, DC: Government Printing Office, 1980.

U.S. National Aeronautics and Space Administration. *Access*. (16mm. Color. 23 min.). 1975 (Film ID No. HQ241). Available on free loan from NASA Regional Film Libraries at NASA centers throughout the U.S.

U.S. National Aeronautics and Space Administration. Scientific and Technical Information Branch. *The NASA Information System and How to Use It.* Washington, DC: NASA, n.d. (Copies are available from issuing agency).

U.S. National Center for Health Statistics. *Catalog of Public Use Data Tapes from the National Center for Health Statistics.* (DHHS Pub. No. PHS 81-1213). Hyattsville, MD: U.S. Department of Health and Human Services, 1980 (HE20.6202:D26).

OTHER RESOURCES

Adams, Margaret O'Neill. "Online Numeric Data-Base Systems: A Resource for the Traditional Library." *Library Trends* 30 (Winter 1982): 435-54.

Adkinson, Burton W. *Two Centuries of Federal Information.* Stroudsburg, PA: Dowden, Hutchinson, & Ross, Inc., 1978.

Berninger, Douglas E., and Burton W. Adkinson. "Interaction between the Public and Private Sectors in National Information Programs." In Martha E. Williams, ed., *Annual Review of Information Science and Technology*, vol. 13. White Plains, NY: Knowledge Industry Publications, Inc., 1978.

Berry, Joseph K. "Spatial Information Systems: 'Instant' Maps for Analyzing Natural Resources Data." *Special Libraries* 72 (July 1981): 261-69.

"Databases Online." *Online Review* 4 (1980): 101-15.

Hunt, Deborah S. "Accessing Federal Government Documents Online." *Database* 5 (February 1982): 10-17.

Luedke, James A., Gabor J. Kovacs, and John B. Fried. "Numeric Data Bases and Systems." In Martha E. Williams, ed., *Annual Review of Information Science and Technology*, vol. 12. White Plains, NY: Knowledge Industry Publications, Inc., 1977.

Luedke, James A., Jr. "Numeric Data Bases On-Line." *On-Line Review* 1 (September 1977): 207-15.

McCarn, Davis B. "Online Systems – Techniques and Services." In Martha E. Williams, ed., *Annual Review of Information Science and Technology*, vol. 13. White Plains, NY: Knowledge Industry Publications, Inc., 1978.

Nakata, Yuri. *From Press to People: Collecting and Using U.S. Government Publications.* Chicago: American Library Association, 1979.

Rowe, Judith S. "Machine-Readable Files of Government Publications." *Government Publications Review* 5 (1978): 195-97.

Rowe, Judith S., and Constance F. Citro. "Machine-Readable Data Files of Government Publications: Data Files from the National Center for Health Statistics." *Government Publications Review* 8A (1981): 135-40.

Rowe, Judith S. "Machine-Readable Data Files of Government Publications: New Sources of Machine-Readable Data on Drug Use, Crime, and Aging." *Government Publications Review* 7A (1980): 417-21.

Williams, Martha E. "Database and Online Statistics for 1979." *ASIS Bulletin* 7 (December 1980): 27-29.

12

INFORMATION ANALYSIS CENTERS

The enormous growth of scientific literature after the Second World War and the resulting difficulty in keeping up with and evaluating the published literature encouraged the growth of formalized information analysis centers.[1] An information analysis center, according to the panel on Information Analysis Centers of the Committee on Scientific and Technical Information (COSATI) is

> ... a formally structured organizational unit specifically (but not necessarily exclusively) established for the purpose of acquiring, selecting, storing, retrieving, evaluating, analyzing, and synthesizing a body of information and/or data in a clearly defined specialized field or pertaining to a specific mission with the intent of compiling, digesting, repackaging, or otherwise organizing and presenting pertinent information and/or data in a form most authoritative, timely, and useful to a society of peers and management.[2]

The rationale behind the information analysis center is that a researcher or an engineer has a need for ready access to reliable information rather than citations to published literature.[3]

The information analysis centers are staffed largely by scientists and, according to Kertesz, are rooted in the tradition of the nineteenth-century scientists such as Beilstein and Gmelin "who accepted the challenge to bring some kind of order into the ever-increasing flood of data, to make experimental findings conveniently available to other scientists."[4] The information analysis centers differ from libraries in that they not only perform document handling functions such as collecting, cataloging, and abstracting but also perform information handling activities such as analyzing, evaluating and synthesizing the published information.[5]

Currently there are more than one hundred federally supported information analysis centers, most of which are in the areas of science and technology.[6] Many of these centers are supported by agencies such as the Department of Defense and the Department of Energy. Many others are part of the National Standard Reference Data System (NSRDS), which is administered by the National Bureau of Standards. The NSRDS, national in scope, was established in 1963 to coordinate the production and dissemination of critically evaluated data in physical sciences.

The information analysis centers provide a variety of services to the scientific community, such as publication of critical data compilations, state-of-the-art reviews, bibliographies, current awareness services, and newsletters. Many centers answer inquiries and provide consulting and literature-searching services.

Bibliographic products centers are available through normal channels such as the Government Printing Office and the National Technical Information Service. For example, a number of data compilations issued by the National Standard Reference Data System (NSRDS) are available through the Superintendent of Documents as NSRDS-NBS Series (C13.48:(nos.)).

A number of other NSRDS publications, including those available in computer-readable magnetic tape format, are available from the National Technical Information Service. In addition, NSRDS data compilations and related publications are made available through channels outside of GPO and NTIS. For example, NSRDS uses the *Journal of Physical and Chemical Reference Data* as its principal publication outlet for the critically evaluated data.

Similar practice is followed by the Nuclear Data Center at Brookhaven National Laboratory in publishing its data compilations and bibliographies in *Nuclear Data Sheets*, a periodical published by Academic Press. Finally, numerous NSRDS publications are available from commercial publishers as well; e.g., Paul Duby, *Thermodynamic Properties of Copper-Slay Systems* (New York: International Copper Research Association, 1976). The information analysis centers, their services and products represent a complex national and international scientific information network.

SOURCES OF INFORMATION ON IACs

Sources of information on information analysis centers (IACs) include *Directory of Federally Supported Information Analysis Centers*, 4th ed. (Washington, DC: Government Printing Office, 1979, LC1.31:In3/979). This directory lists 108 centers, each described in terms of its mission, scope, holdings, publications, services, and the type of users served. Also included about each center is its address, telephone number, name of the director of the center, federal and non-federal organizations that sponsor the center, date the center was established and the number and general qualifications of the staff. The directory has personal name, geographic, sponsors and other organizations, and subject indexes.

Another useful source is the *Encyclopedia of Information Systems and Services*, 5th ed. (Detroit: Gale Research Co., 1982) which provides access to information analysis centers through one of its indexes called Data Collection and Analysis. Information on each center listed in the encyclopedia includes address of the center, contact person, description of the system or service, scope, publications, and computer-based products and services.

Department of Defense Information Analysis Centers: Profiles for Specialized Technical Information (U.S. Defense Technical Information Center, Washington, DC: Government Printing Office, 1981) lists twenty centers, some of which are not listed in the National Referral Center's *Directory*. The profiles describe each center under the headings: address, director, service point of contact, technical monitor, and subject coverage. Another source in this regard, *Defense Technical Information Center (DTIC) Referral Data Bank Directory*, 8th ed. (Alexandria, VA: DTIC, 1981, DTIC/TR-81/1), also lists a number of information analysis centers. This directory provides information on each center under the headings: director and/or contacts, controlling organization/type of source, language input, availability/service charges, descriptors, services/materials, publications, and annotation.

Similarly, *Resource Directory of DOE Information Organizations* (Washington, DC: Government Printing Office, 1981, DOE/TIC-4616) provides a narrative description of several DOE-supported information analysis centers. *Special Technology Groups: 1980 Catalog* (Springfield, VA: NTIS, 1980) describes eighteen information analysis centers and lists their publications. These publications are available from NTIS.

Finally, *International Compendium of Numerical Data Projects: A Survey and Analysis* (International Council of Scientific Unions. Committee on Data for Science and Technology, New York: Springer-Verlag, 1969) and *CODATA Bulletin* place the activities of the federally supported information analysis centers within the context of the worldwide scientific communication network.

The following is a list of selected information analysis centers and descriptions of their products and services.

Name: **Aerospace Structures Information and Analysis Center**
Address: AFWAL/FIBR/ASIAC, Wright-Patterson AFB, OH 45433
Telephone: (513)252-1630; Autovon: 785-6688
Eligibility: U.S. government agencies and contractors
Scope: Structural analysis, fracture mechanics and analysis, fatigue, experimental structural data, etc.
Products & Services:

- Bibliographic searches of DTIC, NTIS, NASA and other data bases
- Distribution of computer programs
- Performing structural analysis and testing
- *ASIAC Newsletter*

For Further Information: *ASIAC: Aerospace Structures Information and Analysis Center*

Name: **Biomedical Computing Technology Information Center**
Address: R-1302, Vanderbilt Medical Center, Nashville, TN 37232

Telephone: (615)322-2385
Sponsor: Department of Energy
Scope: Emphasis is on nuclear medicine computing technology. Seeks to improve the effectiveness of use of computers in biomedical research and development and medical clinical practice.
Products & Services:

- Acts as a clearinghouse for algorithms, documentation and computer programs
- Publishes *BCTIC Newsletter*
- Conducts seminars, workshops and meetings on biomedical computer technology and publishes the proceedings

Name: **Chemical Propulsion Information Agency**
Address: CPIA, The Johns Hopkins University, Applied Physics Laboratory, Laurel, MD 20810
Telephone: (301)953-7100, ext. 7800
Sponsors: Defense Logistics Agency, Army, Navy, NASA, Air Force, and other government agencies
Eligibility: Must be registered with Defense Technical Information Center at least at confidential level and have a "need-to-know"
Scope: Solid and liquid propellants, air breathing propulsion, chemical combustion, rocket motor and nozzle technology
Products:

- Publications
  The Chemical Propulsion Abstracts (4 to 6 times/year) (confidential publication)
  The CPIA Bulletin (bimonthly newsletter) (unclassified)
  Chemical Propulsion Technology Reviews (semiannual)
  CPIA Propulsion Manuals (Rocket Motor Manual, Solid Propellant Manual, Propellant Ingredient Manual, Liquid Propellant Manual, Liquid Propellant Engine Manual, and Airbreathing Propulsion Manual)
  Technical papers presented at annual JANNAF propulsion meetings and other meetings
  Publications of JANNAF (Joint Army-Navy-NASA-Air Force) subcommittees
  An annual list of R&D programs sponsored by JANNAF agencies
  A semiannual listing of selected bibliographies, handbooks, manuals and reviews
- Published literature searches
- Technical and bibliographic inquiry services
- A library of more than fifty thousand publications

For Further Information:

- *Chemical Propulsion Information Agency* (brochure)
- *Basic Services Available to the CPIA, 1981-1982* (brochure)
- *CPIA Technical Inquiries and Literature Searches*
- *Joint Army-Navy-NASA-Air Force Interagency Propulsion Committee* (1982 edition)

Name: **Concrete Technology Information Analysis Center (CTIAC)**
Address: Waterways Experiment Station, Corps of Engineers, Department of the Army, P.O. Box 631, Vicksburg, MS 39180
Telephone: (601)634-3264; FTS 542-3264
Sponsor: Office, Chief of Engineers and U.S. Army Development and Readiness Command (DARCOM)
Scope: Concrete technology including concrete, cement, fire resistance, earthquake effects, etc.
Products & Services:

- Compilation of bibliographies
- Consultation
- Data compilation
- Indexing
- Literature survey
- Referral
- State-of-the-art studies
- Technical analysis and evaluation
- Technical inquiries

Related Information Analysis Centers:
Hydraulic Engineering Information
Analysis Center (HEIAC), Pavements and Soil
Trafficability Information Analysis Center (PSTIAC) and
Soil Mechanics Information Analysis Center (SMIAC)
These are located at the same address above.
For Further Information:
Chief, Technical Information Center
Attn: WESTV, USAEWES
P.O. Box 631
Vicksburg, MS 39180
Telephone: (601)634-2533; FTS 542-2533

Name: **Controlled Fusion Atomic Data Center**
Address: Physics Division, Oak Ridge National Laboratory, Oak Ridge, TN 37830
Telephone: (615)574-4704; FTS 624-4704
Sponsor: Office of Fusion Energy, Department of Energy
Scope: Numerical atomic collision data of interest to controlled thermonuclear reactions
Products & Services:

- Answers inquiries
- Publishes *Atomic Data for Fusion* (bimonthly), a newsletter published jointly with Atomic Transitioin Probabilities center (National Bureau of Standards)
- *Atomic Data for Controlled Fusion Research* (1977) (2v) (Third volume is in progress)
- *Bibliography of Atomic and Molecular Processes*

For Further Information:
Controlled Fusion Atomic Data Center (March 1982)

Name: **Data & Analysis Center for Software**
Address: RADC/ISISI, Griffiss AFB, NY 13441
Telephone: (315)336-0937; Autovon: 587-3395
Sponsor: Air Force Systems Command, Rome Air Force Development Center; operated by IIT Research Institute
Scope: DACS, a DoD information analysis center, acts as a centralized source of information on computer software technology
Products:

- Software Life Cycle Empirical Database (SLED) and related products. (This data base consists of software experience data such as Software Problem Reports. Subsets of this data base are available in hardcopy or on magnetic tape.)
- Bibliographic Products, e.g., DACS Annotated Bibliography and Directory of Software Engineering Research Products
- Computer Searches on DACS Bibliographic Data Base
- Data Collection Forms
- State-of-the-Art Reports, e.g., A Review of Software Maintenance Technology
- *DACS Newsletter* (Quarterly)

Product Availability: DACS
For Further Information:

- Data & Analysis Center for Software: Products & Services Information
- *Users' Guide to the DACS Products & Services* (February 1982)
- *Users' Guide to Bibliographic Services: Custom Searches* (February 1982)

Name: **Data Center on Atomic Transition Probabilities**
Address: National Bureau of Standards Building, 221, A267, Washington, DC 20234
Telephone: (301)921-2071; FTS: 921-2071
Director: Dr. Wolfgang L. Wiese
Contacts: Georgia A. Martin and Jeffrey R. Fuhr
Scope: Transition Probabilities of Atoms and Atomic Ions in the Gas Phase
Products:

- Critically evaluated data, e.g., J. R. Fuhr et al., "Atomic Transition Probabilities for Iron, Cobalt, and Nickel (A Critical Data Compilation of Allowed Lines)," *Journal of Physical and Chemical Reference Data* 10 (1981):305-565.
- Bibliographies, e.g., B. J. Millar, J. R. Fuhr, and G. A. Martin, *Bibliography on Atomic Transition Probabilities (November 1977 through March 1980)*, NBS-Spec. Publ. 505, Suppl. 1, Washington, DC: GPO, 1980.
- Critical reviews, e.g., W. L. Wiese, "Atomic Transition Probabilities and Lifetimes," in W. Hanle and H. Kleinpoppen, eds., *Progress in Atomic Spectroscopy*, Part B, Ch. 25, Plenum Publishing Corp., 1979, pp. 1101-55.

Product Availability: Government Printing Office and professional journals
For Further Information:
Complete bibliography available from the center

Name: **DoD Nuclear Information and Analysis Center (DASIAC)**
Address: Kaman Tempo - DASIAC, 816 State Street, P.O. Drawer QQ, Santa Barbara, CA 93102
Telephone: (805)963-6444-453 (general information)
Sponsor: Defense Nuclear Agency
Eligibility: DNA personnel and DNA contractors: DoD and other federal agencies' personnel and contractors
Scope: Strategic and tactical nuclear warfare, nuclear test detection, nuclear weapon safety and security, survivability and vulnerability, phenomena associated with fireball, nuclear weapon output, etc.
Products & Services:

- Collection and preservation of raw and reduced test data
  High-speed optical records
  Data in the form of film, magnetic tape, scope recordings
  Project documentation such as project notebooks and logs
- Dissemination of experimental data and computer codes that calculate shock characteristics, etc.
- Document acquisition and dissemination
- Acquisition and dissemination of computer codes
- Data analysis
  Surveys, reviews, nuclear test and simulation data compilations, literature guides, indexes, current awareness listings, and special bibliographies
- Preparation and editing of DNA handbooks, sourcebooks and manuals
- Meetings, conference, and workshop support

For Further Information:
DASIAC User's Guide and DASIAC References

Name: **Infrared Information and Analysis Center**
Address: Environmental Research Institute of Michigan, P.O. Box 8618, Ann Arbor, MI 48107
Telephone: (313)994-1200, ext. 214
Director: George J. Zissis
Sponsor: Office of Naval Research
Eligibility: DoD contractors on an annual subscription basis
Scope: A DoD information analysis center dealing with the military applications of infrared technology such as the study of radiation characteristics of natural and man-made objects
Products:

- *The Infrared Handbook*
- Assists ONR in conducting the annual Infrared Information Symposium (IRIS) and the meetings of the specialty groups of IRIS
- Assists ONR in conducting DoD Laser Conferences
- Publication of proceedings and minutes of the meetings
- Answers inquiries
- Maintains a computer-based bibliographic information retrieval system

Product Availability: Unclassified publications available from the center and NTIS

Name: **Metals and Ceramics Information Center (MCIC)**
Address: P.O. Box 8128, Columbus, OH 43201
Telephone: (614)424-5000 (general information)
Sponsor: Department of Defense
Scope: Characteristics and utilization of advanced metals, e.g., titanium, aluminum and magnesium, and ceramics, e.g., borides, carbides, and nitrides
Products & Services:

- Current awareness service
  publication of *Current Awareness Bulletin* (*CAB*) (monthly) (free to qualified subscribers)
- Special studies undertaken in response to specific needs of users (e.g., indepth review and analysis of literature on a specific area)
- Answers technical inquiries
- Publications
  Reviews and reports (e.g., *Corrosion Fatigue of Metals in Marine Environments*)
  Bibliographies (e.g., *The Effect of Rapid Heating on the Properties of Materials: A Bibliography with Descriptors, 1948-1976*)
  Proceedings of conferences and symposia (e.g., *Physical Metallurgy of Uranium Alloys*)
  Handbooks (e.g., *Soviet Alloy Handbook, Aerospace Structural Metals Handbook* and *Structural Alloys Handbook*)
  Man Tech Journal (Quarterly)

For Further Information:
Metals and Ceramics Information Center: User's Guide and Materials Information Publications List

Name: **National Nuclear Data Center**
Address: Building 197D, Brookhaven National Laboratory, Upton, NY 11973
Telephone: (516)282-2902
Sponsor: Department of Energy
Eligibility: Users in the U.S. and Canada
Scope: Low energy nuclear physics; specifically, neutron, charged-particle and photonuclear reactions, nuclear structure and decay data
Products & Services:

- Maintenance of bibliographic files
  Neutron Data File (CINDA)
  Charged Particle Data File (CPBIB)
  Nuclear Structure and Radioactive Decay Data File (NSR)
- Maintenance of experimental data file
  Neutron, Charged-particle and Photonuclear Reaction Data (CSISRS)
  Nuclear Structure and Radioactive Decay Data
- Maintenance of evaluated nuclear data
  Neutron data (ENDF/B and ENDF/A files)
  Nuclear Structure Data (ENSDF)
  Radioactive Decay Data
  Charged Particle Nuclear Data
- Retrievals from comptuerized files of CSISRS, ENDF/B, ENDF/A, ENSDF, CINDA, CPBIB, and NSR

- Publications
  NNDC Newsletter
  Neutron Cross Sections (BNL-325)
  Angular Distributions for Neutron-Induced Reactions (BNL-400)
  Bibliography of Integral Charged-particle Nuclear Data (BNL-NCS-50640)
  CINDA
  Nuclear data sheets
  Technical reports
- Coordinates the activities of Cross Section Evaluation Working Group (CSEWG)

For Further Information:
National Nuclear Data Center: Products and Services (March 1982)

Name: **Nondestructive Testing Information Analysis Center (NTIAC)**
Address: Southwest Research Institute, P.O. Drawer 28510, San Antonio, TX 78284
Telephone: (512)684-5111, ext. 2362
Sponsor: Department of Defense
Scope: Nondestructive testing methodology and instrumentation
Products & Services:

- Answers technical inquiries
- Bibliographic services
  conducts literature searches on its own bibliographic file, NTIS file, DTIC files, and other commercially available files
- Current awareness service
  participates in nondestructive testing meetings
  publishes *NTIAC Newsletter*
- Special services
  preparation of major literature and state-of-the-art surveys
  preparation of handbooks, databooks and feasibility studies
  conducting workshops and conferences and publishing their proceedings

For Further Information:
NTIAC (Nondestructive Testing Information Analysis Center) User's Guide

Name: **Nuclear Safety Information Center**
Address: Oak Ridge National Laboratory, P.O. box Y, Oak Ridge, TN 37830
Telephone: (615)574-0391; FTS 624-0391
Sponsor: Department of Energy
Eligibility: Government organizations and their prime contractors only
Scope: Disseminates information on the safe operation of all types of nuclear facilities
Products & Services:

- Publications
  Nuclear Safety: Technical Progress Review
  Bibliographies
  State-of-the-Art Reports
- Technical inquiries
- Retrospective searches on NSC Data Base
- SDI services

Name: **Plastics Technical Evaluation Center (PLASTEC)**
Address: U.S. Army Armament Research and Development Command (ARRADCOM), Dover, NJ 07801
Telephone: (201)328-4222
Eligibility: Available to scientific and technical community on a fee basis
Scope: Plastics, composites, and adhesives
Products & Services:

- Maintenance of COMPAT, a computerized information retrieval system dealing with the effects of explosive materials on polymers. A supplemental program HAZARD-FAILURE provides access to information on known deficiencies of polymers.
- Maintenance of a library of 30,000 documents
- Provides access to its collection of standards and specifications on plastics
- Assists in design and development of computerized files
- Conducts literature searches on request. Has access to DTIC, DIALOG, and ORBIT.
- Publishes technical reports and notes. These are available from NTIS.

For Further Information:
1) PLASTEC: Plastics Technical Evaluation Center
2) A Cumulative Listing: Citation Abstract and Procurement of PLASTEC publications

Name: **Primate Information Center**
Address: Regional Primate Research Center (SJ-50), University of Washington, Seattle, WA 98195
Telephone: (206)543-4376
Sponsor: In part, Animal Resources Program, Division of Research Resources, National Institutes of Health
Scope: Primatological, biomedical, veterinary and psychological studies on nonhuman primates such as monkeys and apes
Products & Services:

- Publications
  Current Primate References (Monthly)
  numerous bibliographies under the general topics of: General, Behavior, Biology and Experimental Biomedicine, Naturally Occurring Diseases, Experimental Techniques, Pharmacology & Toxicology, and Reproduction
- Monthly custom bibliographies
- Retrospective custom searches
- PIC Indexing Vocabulary: Alphabetic List and Hierarchial List

For Further Information:
Primate Information Center

Name: **Radiation Chemistry Data Center**
Address: Radiation Laboratory, University of Notre Dame, Notre Dame, IN 46556
Telephone: (219)283-6527
Sponsors: Office of Standard Reference Data, National Bureau of Standards, and Department of Energy

Scope: Radiation chemistry and photochemistry with emphasis on kinetics and other properties of radicals and other unstable chemical species

Products:

- Publications

  Biweekly List of Papers on Radiation Chemistry and Photochemistry (a current awareness service)

  Data compilations available from the Government Printing Office, issued in NSRDS-NBS Series, e.g., A. B. Ross and P. Neta, *Rate Constants for Reactions of Inorganic Radicals in Aqueous Solution*, Washington: GPO, 1979

  Thesaurus for Radiation Chemistry (PB 267-655)
- Computer Search Services

For Further Information:

Kinetics of Radical Processes in Aqueous Solution: Data Compilations in the NSRDS-NBS Series (Brochure)

W. Phillip Helman and Alberta B. Ross, "Radiation Chemistry Data Center: Information Services Produced from the Bibliographic Data Base," *Radiation Physics and Chemistry* 16 (1980): 425-30.

Name: **Radiation Shielding Information Center (RSIC)**

Address: Oak Ridge National Laboratory, P.O. Box X, Oak Ridge, TN 37830

Telephone: (615)574-6176; FTS 624-6176

Sponsors: Department of Energy, Defense Nuclear Agency, and Nuclear Regulatory Commission

Scope: Radiation protection and radiation transport shielding information. Deals with radiation from reactors, weapons, accelerators and space. Emphasizes aspects such as the study of interaction between radiation and matter, and study of the properties of shielding materials.

Products & Services:

- Answers technical inquiries
- Performs literature searches
- Sponsors workshops and seminars
- Publishes bibliographies and state-of-the-art reports
- Publishes *RSIC Newsletter*
- Packages, distributes, and provides information on computer codes and machine-readable data libraries
- Maintains an information retrieval system
- Maintains an archival microfiche file of analyzed literature
- Participates in computing standards activities
- Participates in Cross Section Evaluation Working Group (CSEWG) activities

For Further Information:

Radiation Shielding Information Center (October 1980)

Name: **Reliability Analysis Center**

Address: Griffiss AFB, NY 13441

Telephone: (315)330-4151; Autovon: 587-4151

Scope: Electronic component reliability data emphasizing microelectronic devices, high-technology, discrete semiconductors and non-electronic parts. Specific interests are failure modes and mechanisms; material, device and process technology; quality assurance; reliability and maintainability practices; specifications and standards; etc.

Products & Services:

- Publications
  Component and equipment databooks (e.g., *Hybrid Circuit Data*, 1980)
  Handbooks (e.g., *Reliability Design Handbook*, March 1976)
  Technical reliability studies (e.g., *ESD Protective Materials and Equipment: A Critical Review*, May 1982)
  Symposia proceedings (e.g., *Electrical Overstress/Electrostatic Discharge, 1980 Symposium Proceedings*)
  RAC Newsletter (Quarterly)
- Reference service including the analysis, evaluation and summarization of the information
- Consulting services
- Conducting symposia and tutorial courses (e.g., Reliability Design Training Course)

Name: **Shock and Vibration Information Center**
Address: SVIC, Naval Research Laboratory, Code 5804, Washington, DC 20375
Telephone: (202)767-2220
Sponsor: Department of Defense
Scope: Shock, vibration and accoustics phenomena in various environments such as bridges and buildings
Products & Services:

- Publications
  Bulletins (Annual) (collection of papers presented at Shock and Vibration Symposia)
  The Shock and Vibration Digest (Monthly)
  The Shock and Vibration Monographs, e.g., R. R. Bouche, *Calibration of Shock and Vibration Measuring Transducers* (SVM-11) (1979)
  Special publications, e.g., C. T. Morrow, *The Environmental Qualification Specification as a Technical Management Tool* (November 1981)
  Catalog of Technical Publications
- Shock and Vibration Information Base
- Conducts workshops and symposia
- Answers technical queries

Name: **Tactical Weapon Guidance & Control Information Analysis Center (GACIAC)**
Address: GACIAC, ITT Research Institute, 10 West Thirty-fifth Street, Chicago, IL 60616
Telephone: (312)567-4519
Eligibility: Industry and government/military users who are registered with Defense Technical Information Center
Scope: Tactical weapon guidance and control systems. Tactical weapons of interest are missiles, rockets, bombs, projectiles, and munition-dispensing cannisters.
Products & Services:

- Abstracts and indices

- Critical reviews and technology assessments
- The *GACIAC Bulletin*
- Handbooks and databooks
- Proceedings of conferences/workshops
- Reports of special studies and tasks
- State-of-the-art reviews
- Answers bibliographic and technical inquiries

For Further Information:
GACIAC Users' Guide and Participation Plan

Name: **Thermophysical and Electronics Properties Information Analysis Center**
Address: TEPIAC/CINDAS, Purdue University, 2595 Yeager Road, West Lafayette, IN 47906
Telephone: (317)494-6300
Scope: Thermophysical, electrical and magnetic and other properties of materials
Products & Services:

- Maintenance of computerized Material Properties Bibliographic Data System
- Maintenance of computerized Material Properties Numerical Data System
- Evaluation of numerical data to obtain recommended reliable values
- Technical consulting
- Retrospective bibliographic searches
- Document reproduction services
- Publications
  Thermophysical Properties of Matter: The TPRC Data Series (13v)
  Thermophysical Properties Research Literature Retrieval Guide (3v) Supplement I [6v]) and (Supplement II [6v]) plus a number of other publications available from Plenum Publishing Corporation. Some other publications are available from McGraw-Hill Book Company.

For Further Information:
Thermophysics and Electronics Newsletter 11 (March 1982)

Name: **Toxicology Information Response Center**
Address: Oak Ridge National Laboratory, P.O. Box X, Oak Ridge, TN 37830
Telephone: (615)576-1743
Sponsor: National Library of Medicine's Toxicology Information Program
Scope: TIRC is concerned with toxicological effects of chemicals and chemical classes that occur in foods and environment. It is a part of Information Center Complex (ICC) of the Oak Ridge National Laboratory. ICC encompasses, in addition to TIRC, the following centers: Carbon Dioxide Information Center, Chemical Effects Information Center, Center for Energy and Environmental Information, Ecological Sciences Information Center, Environmental Mutagen Information Center, Environmental Teratology Information Center, Fossil Energy Center, Hazardous Materials Information Center, and Toxicology Data Bank.
Products & Services:

- Individualized literature searches–searches are conducted on data bases such as MEDLINE, TOXLINE, RECON, and the data bases available on DIALOG, ORBIT, and BRS services

- Selective dissemination of information
- Bibliographies—available from NTIS and Federation of American Societies for Experimental Biology, Bethesda, MD 20014
- Specific toxicology data and state-of-the-art reviews

Name: **X-Ray and Ionizing Radiation Data Center**
Address: Dr. John H. Hubbell, Radiation Physics Division, National Bureau of Standards, Washington, DC 20234
Scope: It is one of the NSRDS (National Standard Reference Data Systems) centers. Most of these centers are located at NBS. X-Ray and Ionizing Radiation Data Center is concerned with collecting and evaluating x-ray attenuation and related data.
Products & Services:

- Publications through NBS and non-NBS channels. One of the center's publications, NSRDS-NBS 29, became a citation classic. Data published in sources such as *Journal of Physical and Chemical Reference Data*, *Radiation Research*, and *International Journal of Applied Radiation and Isotopes*
- Answers technical inquiries
- Provides consulting services

NSRDS Data Centers:

Atomic Collision Information Center, Joint Institute for Laboratory Astrophysics, University of Colorado, Boulder, CO 80309

Center for Information and Numerical Data Analysis and Synthesis, Purdue University, 2595 Yeager Road, West Lafayette, IN 47906

JANAF Thermochemical Tables, Dow Chemical Company, 1707 Building, Thermal Research Laboratory, Midland, MI 48640

Molten Salts Data Center, Rensselaer Polytechnic Institute, Department of Chemistry, Troy, NY 12181

National Center for Thermodynamic Data of Minerals, U.S. Geological Survey, 959 National Center, Reston, VA 22092

Radiation Chemistry Data Center, Radiation Laboratory, University of Notre Dame, Notre Dame, IN 46556

Thermodynamics Research Center, Department of Chemistry, Texas A&M University, College Station, TX 77843

Thermodynamic Research Laboratory, Department of Chemical Engineering, Washington University, St. Louis, MO 63130

Centers Located at NBS:

Alloy Phase Diagram Data Center
Aqueous Electrolyte Data Center
Atomic Energy Levels Data Center
Atomic Transition Probabilities and Atomic Line Shapes and Shifts Data Center
Chemical Kinetics Information Center
Chemical Thermodynamics Data Center
Crystal Data Center
Diffusion in Metals Data Center
Fluid Mixtures Data Center

Fundamental Constants Data Center
Ion Energetics Data Center
Molecular Spectra Data Center
Phase Diagrams for Ceramists Data Center
Photon and Charged-Particle Data Center
Photonuclear Data Center
X-Ray and Ionizing Radiation Data Center

Other Centers (not funded or administered by the Office of Standard Reference Data:

Fundamental Particle Data Center, Lawrence Berkeley Laboratory, University of California, Berkeley, CA 94720

High Pressure Data Center, P.O. Box 7246, University Station, Provo, Utah 84602

Isotopes Project, Lawrence Berkeley Laboratory, University of California, Berkeley, CA 94720

Source: John D. Hoffman and David R. Lide, *Office of Standard Reference Data: Annual Report 1982; NAS-NRC Evaluation Panel, November 15-16, 1982*, Washington, DC: National Bureau of Standards, 1982, Appendix H: NSRDS Data Centers, pp. 40-44.

REFERENCES

1. A recent review of literature on information analysis centers is Bonnie (Talmi) Carroll and Betty F. Maskewitz, "Information Analysis Centers," *Annual Review of Information Science and Technology* 15 (1980): 147-89. See also Herman M. Weisman, *Information Systems, Services, and Centers* (New York: Wiley, 1972), chapters 9-11; and Stephen A. Rossmassler and David G. Watson, eds., *Data Handling for Science and Technology: An Overview and Sourcebook* (Amsterdam: North-Holland, 1980).

2. U.S. Federal Council for Science and Technology, Committee on Scientific and Technical Information, Panel 6, Information Analysis and Data Centers, *Directory of Federally Supported Information Analysis Centers* (COSATI-70-1) (PB 189-300) (Washington, DC: U.S. Government Printing Office, 1970), p. iii.

3. United States President's Science Advisory Committee, *Science, Government, and Information: The Responsibilities of the Technical Community and the Government in the Transfer of Information: A Report* (Washington, DC: U.S. Government Printing Office, 1963), p. 32 (Weinberg Report).

4. Francois Kertesz, "Collaboration between Information Analysis Centers at a Large Multipurpose Library," in *Handling of Nuclear Information: Proceedings of a Symposium, Vienna, 16-20 February 1970* (Vienna, Austria: International Atomic Energy Agency, 1970), pp. 103-10 (IAEA-SM-128/290). The Weinberg Report itself (U.S. President's Science Advisory Committee, op. cit., p. 33) draws upon the experience of Nuclear Data Center whose information analysis activities "contributed notably, for example, to the development of the Shell Model of the nucleus, one of the major theoretical underpinings of modern nuclear physics."

5. United States Library of Congress, National Referral Center, *Directory of Federally Supported Information Analysis Centers*, 4th ed. (Washington, DC: U.S. Government Printing Office, 1979), p. iii.

6. Ibid.

13

REFERENCE SOURCES

Numerous scientific and technical reference sources are published by the federal government. Types of reference sources available include bibliographies, dictionaries, glossaries, thesauri, directories, handbooks and manuals, atlases, yearbooks, monographs, and treatises. These sources represent secondary literature in the process of scientific communication.

Bibliographies, along with the previously discussed indexing and abstracting services and current awareness services are surrogates of primary literature; dictionaries, directories, and handbooks and manuals are repackages of primary literature; and monographs and treatises are compact versions of primary literature.[1]

These reference sources provide a ready access to a variety of information such as the meanings of technical terms, addresses of scientific organizations, and specific methods and techniques of, say, chemical analysis of wastewater. Although many government reference sources are highly specialized, they are helpful to researchers and reference librarians as they complement the commercially available reference materials.

The reference sources available through government documents collections are too numerous to be listed in a guide of this type. Many of them can be retrieved easily through the *Monthly Catalog* and *Government Reports Announcements & Index* (*GRA&I*).[2] The recent introduction of a keyword index in the *Monthly Catalog* is helpful in expediting searches for reference sources since searches can be conducted on keywords such as "bibliography," "bibliographies," "directory," "handbook," and the like.

Similarly, searching the keyword indexes of *GRA&I* under similar terms would also retrieve a number of reference tools. In addition, one can also scan the abstracts under the Subject Category Group 5B of each issue. Group 5B represents library and information sciences and many reference materials are listed under this category in the issues of *GRA&I.*

Searching the National Technical Information Service's weekly current awareness publication, *Library & Information Sciences*, would also be helpful since this publication brings together the reference sources under the heading "reference materials."

Statistical compilations relevant to science and technology can be retrieved using *American Statistics Index* (Washington, DC: Congressional Information Service). *ASI*, of course, specifically excludes scientific and technical data but it does include statistical data such as the expenditures of scientific research and development and employment data of scientists and engineers.

Another important source for locating government reference materials is *Government Reference Books* (1968/69-), a biennial publication issued by Libraries Unlimited, Inc. The most recent issue of this source is *Government Reference Books 80/81: A Biennial Guide to U.S. Government Publications* (Littleton, CO: Libraries Unlimited, 1982). This publication limits itself to GPO documents but covers all subject areas. In addition, having access to online files of GPO's *Monthly Catalog*, NTIS, and the Defense Technical Information Center should make it easy to locate appropriate reference materials.

BIBLIOGRAPHIES

A great number of bibliographies are announced through government publication channels. Many of these, of course, are book-length bibliographies. Roberta A. Scull's two-volume work, *A Bibliography of United States Government Bibliographies* (Ann Arbor, MI: Pierian Press, 1975 and 1979), covers the period 1968 through 1974 and lists over twenty-five hundred government-produced bibliographies in all subject areas.

Similarly, the magnitude of bibliographic information one can locate in the documents collections can be seen from Edna A. Kanely's *Cumulative Subject Guide to U.S. Government Bibliographies 1924-1973*, 7v. (Arlington, VA: Carrollton Press, 1976). Associated with this source is U.S. Government Bibliography Masterfile, 1924-1973 which is a microfiche collection of full texts of bibliographies listed in the guide.

Another useful bibliography of bibliographies is Gabor Kovacs' *Annotated Bibliography of Bibliographies on Selected Government Publications and Supplementary Guides to the Superintendent of Documents Classification System*, Sixth Supplement (Greeley: University of Northern Colorado, 1980) and its predecessors prepared by Alexander C. Body. This bibliography of bibliographies consists of extensive annotations for each publication listed.

Kovacs states that the sixth supplement includes more than one thousand and eighty bibliographies available through the federal depository libraries which were published during 1978 and 1979. In any case, the number of bibliographies one can locate through *Monthly Catalog* and *GRA&I* is impressive.

The following are examples of bibliographies available in the government documents collections:

Craven, Scott R., comp. *The Canada Goose (Branta Canadensis): An Annotated Bibliography*. Washington, DC: U.S. Department of the Interior. Fish and Wildlife Service, 1981. 66p.

U.S. Department of Energy. Office of Energy Research. *1980 Bibliography of Atomic and Molecular Processes*. Springfield, VA: National Technical Information Service, 1982. 428p. (DOE/ER-0118).

In addition to these bibliographies, the online availability of sources such as NTIS and MEDLARS give rise to numerous computer-produced bibliographies. NTIS, for example, lists more than thirty-five hundred such bibliographies in its *NTIS Published Searches Master Catalog* (Springfield, VA: NTIS, 1982). Examples of such bibliographies are

Bacterial Pollution of Water. March, 1979-July, 1982. (Citations from the NTIS Data Base). Springfield, VA: National Technical Information Service, 1982. 290p. (PB 82-810680).

Zinc Toxicology. January 1977 through July 1982. (NLM Literature Search No. 82-10). Bethesda, MD: National Library of Medicine. 204 citations.

The Government Printing Office publishes a series of bibliographies on a variety of topics. These bibliographies, numbering over two hundred and fifty and called Subject Bibliographies, cover many scientific and engineering subjects. These are free and a recent list of Subject Bibliographies can be requested from the U.S. Government Printing Office, Superintendent of Documents, Washington, DC 20402. They also can be located in many depository libraries. The following are examples of GPO Subject Bibliographies in the area of science/technology:

Agricultural Research, Statistics, and Economic Reports	SB 162
Agriculture Yearbooks (Department of)	SB 31
Air Pollution	SB 46
Aircraft, Airports, and Airways	SB 13
Astronomy and Astrophysics	SB 115
Computers and Data Processing	SB 51
Fish and Marine Life	SB 209
Insects	SB 34
Mathematics	SB 24
Noise Abatement	SB 63
Oil Spills and Ocean Dumping	SB 79
Science Experiments and Projects	SB 243
Smoking	SB 15
Space, Rockets, and Satellites	SB 297
Telecommunications	SB 296
Wildlife Management	SB 116
X-Rays	SB 193
Zoology	SB 124

The Government Printing Office periodically issues an index to these Subject Bibliographies.

ATLASES

Atlases of trees, bird habitats, planets and oceanography are available in the documents collections. These sources provide pictorial representation of the topics they cover, e.g., geographical distribution of certain types of trees or illustrations of anatomical parts of humans and animals. The following are some recent examples of such sources:

McMillin, Charles W. *The Wood and Bark of Hardwoods Growing on Southern Pine Sites: A Pictorial Atlas.* Washington, DC: U.S. Government Printing Office, 1980. 58p.

Southern Forest Experiment Station, New Orleans, Louisiana. *A Forest Atlas of the South.* Washington, DC: Department of Agriculture, Forest Service, 1981. 27p.

Batson, Raymond M., et al. *Atlas of Mars.* Washington, DC: National Aeronautics and Space Administration, 1979. 146p.

U.S. Department of the Navy. Naval Oceanography and Meteorology. *U.S. Navy Marine Climatic Atlas of the World.* Washington, DC, 1974-81.

Antonovych, T. T. *Atlas of Kidney Biopsis.* Washington, DC: Armed Forces Institute of Pathology, 1980. 386p.

Mason, Thomas J., et al. *An Atlas of Mortality from Selected Diseases.* (NIH Pub. No. 81-2397). Washington, DC: U.S. Government Printing Office, May 1981.

Portnoy, John W., et al. *Atlas of Gull and Tern Colonies: North Carolina to Key West, Florida* (Including Pelicans, Cormorants, and Skimmers). Washington, DC: Department of the Interior, Fish and Wildlife Service, National Coastal Ecosystems Team, 1981.

Scully, Robert F. *Tumors of the Ovary and Maldeveloped Gonads: Atlas of Tumor Pathology.* Washington, DC: Armed Forces Institute of Pathology, 1979. 413p.

GLOSSARIES

Federal publications include a variety of dictionaries and glossaries in many specialized fields of science and technology. These, of course, complement the standard dictionaries and glossaries published by commercial publishers. Glossaries and dictionaries available from the federal agencies are particularly useful since they may be dealing with relatively recent areas of science and technology. Examples of such publications are

Hanchett, J. G., and F. W. Hasselberg. *Glossary of Terms.* Washington, DC: U.S. Nuclear Regulatory Commission, Office of Public Affairs, 1981. 52p.

Kaye, Seymour M. *Encyclopedia of Explosives and Related Items.* Dover, NJ: U.S. Army Armament Research and Development Command, Large Caliber Weapon Systems Laboratory, 1978.

Hanson, A. G., et al. *Optical Waveguide Communications Glossary.* (NTIA-Sp-79-4). Washington, DC: Department of Commerce, National Telecommunications and Information Administration, 1979.

U.S. Air Force. Air Intelligence Group (7602nd). *Chinese-English Rocketry Dictionary*, v. 3. (AD-AO85 074/3). San Francisco, 1979. 817p. (Available: NTIS).

ACTION/Peace Corps. Information Collection and Exchange. *A French/English Glossary of Agricultural Terms.* (PB 80-159197). Washington, DC, 1979. 64p. (Available: NTIS).

Matthews, John E. *Glossary of Aquatic Ecological Terms.* (AD-AO91 048). Ada, OK: Robert S. Kerr Environmental Research Laboratory, 1972. (Available: NTIS).

Advisory Group for Aerospace Research and Development, Neuilly-sur-Seine (France). *Multilingual Aeronautical Dictionary (Dictionaire Aeronautique Multilingue).* (AD-AO95 571). Neuilly-sur-Seine, France, 1980. 897p.

DIRECTORIES

Directories of organizations, people, research projects in progress and such others are common among government publications and technical reports. These are useful in locating the addresses and other relevant information about scientists and their organizations. Another useful source for federal directories is Donna Rae Larson's *Guide to U.S. Government Directories, 1970-1980* (Phoenix, AZ: Oryx Press, 1981). The following are examples of directories:

U.S. Department of Transportation. *Federal Aviation FAA Directory*. Washington, DC: U.S. Government Printing Office.

U.S. Department of Health and Human Services. Centers for Disease Control. Center for Prevention Services. Venereal Disease Control Division. *Directory of STD [Sexually Transmitted Diseases] Clinics.* Washington, DC: U.S. Government Printing Office, 1981.

U.S. Department of Health and Human Services. National Institutes of Health. Division of Research Resources. *Biotechnology Resources: A Research Resources Directory*. Washington, DC: U.S. Government Printing Office, 1981.

U.S. Department of Health and Human Services. Public Health Service. Office on Smoking and Health. *1980 Directory [of] Ongoing Research in Smoking and Health*. Washington, DC: U.S. Government Printing Office, 1980.

U.S. Department of Agriculture. Food and Nutrition Service. *Directory of Cooperating Agencies.* Washington, DC: U.S. Government Printing Office, 1981.

U.S. Department of the Interior. *Information Sources and Services Directory.* Washington, DC: Government Printing Office, 1979.

U.S. Department of Health and Human Services. National Institutes of Health. National Cancer Institute. *Directory of Cancer Research: Information Resources.* Washington, DC: U.S. Government Printing Office, 1981 (PB 81-163537).

U.S. Central Intelligence Agency. National Foreign Assessment Center. *Directory of Soviet Research Organizations: A Reference Aid.* Washington, DC: 1978. 290p. (PrEx3.11:CR78-11336).

U.S. Department of Commerce. National Technical Information Service. *A Directory of Computer Software Applications: Civil & Structural Engineering, 1978-Sept 1980.* Springfield, VA, 1981 (C51.11/5:C49/978-80).

U.S. Department of Energy. Office of Energy Research. Division of Nuclear Physics. *Directory of Contractors Supported by Division of Nuclear Physics.* Springfield, VA: National Technical Information Service, 1981 (E1.19:0114) (DOE/ER-0114).

In addition, commercially produced directories also provide access to federal information sources. A number of such directories are published by publishers such as Washington Researchers (918 Sixteenth Street, NW, Washington, DC 20006) and Carroll Publishing Company (1058 Thomas Jefferson Street, NW, Washington, DC 20007). Washington Researchers, for example, publishes directories called *Researcher's Guide to Washington Experts* and *Washington Information Workbook.* The *Experts Directory* is said to provide name, telephone number, and address for each of 15,000 federal bureaucrats with an expertise in 12,000 topics.

The Washington Researchers also publishes a monthly newsletter called *The Information Report* providing updates on the information activities of federal agencies. The Carroll Publishing Company publishes *Federal Executive Directory* which also provides telephone number and agency name of a number of federal personnel. These sources are useful in tracking down federal information.

HANDBOOKS AND MANUALS AND NUMERIC DATA COMPILATIONS

Handbooks and manuals provide specific practical and theoretical information to scientists and engineers. Usually, they are valuable compendia of procedures, chemical and/or mathematical formulae, properties of materials and related numerical data tables which aid the engineers or scientists working in a laboratory. As has been pointed out in the chapters on information analysis centers and standards and specifications, federal scientists are active in compiling evaluated numeric data to make sure that scientists have access to reliable data.

Numerous handbooks and manuals and numeric data compilations are published as federal publications. One useful source in locating numerical data compilations is the U.S. National Bureau of Standards' *Standard Reference Data Publications, 1964-1980*, compiled by Gertrude B. Sherwood (Washington, DC: Government Printing Office, 1981, NBS SP-612, C13.10:612). The following are examples of handbooks and manuals published by the federal agencies:

Federal Bureau of Investigation. *Handbook of Forensic Science*. Washington, DC: U.S. Government Printing Office, 1979.

Miller, James W., ed. *NOAA Diving Manual: Diving for Science and Technology*. 2d ed. Washington, DC: U.S. Government Printing Office, 1979.

Ratliff, Thomas A., et al. *Industrial Hygiene Laboratory Quality Control Manual*. Cincinnati, OH: Department of Health, Education, and Welfare, Center for Disease Control, National Institute for Occupational Safety and Health, July 1979 (HE20.7111/2:In2).

National Library of Medicine. *NLM Online Services Reference Manual* (January 1980), 582p. (PB 80-114531).

U.S. Environmental Protection Agency. *Air Pollution Engineering Manual*. 2d ed. Research Triangle Park, NC, 1973 (EP4.9:40).

McCarty, R. D., J. Hord, and H. M. Roder. *Selected Properties of Hydrogen (Engineering Design Data)*. (NBS Monograph 168). Washington, DC: U.S. Government Printing Office, 1981 (C13.44:168).

Sieck, L. Wayne. *Rate Coefficients for Ion-Molecule Reactions, II. Organic Ions Other Than Those Containing Only C and H*. (NSRDS-NBS 64). Washington, DC: U.S. Government Printing Office, 1979 (C13.48:64).

THESAURI

The existence of a great number of computerized information retrieval systems has given rise to a proliferation of thesauri in the documents collections. These thesauri are of use to science and technology librarians in conducting searches of computerized data bases. Examples of thesauri are

U.S. Health and Human Services. National Institutes of Health. *Medical and Health Related Sciences Thesaurus*. (NIH Pub. No. 80-199). Bethesda, MD, 1980 (HE20.3023:980).

U.S. Department of the Interior. Office of Water Research and Technology. *Water Resources Thesaurus: A Vocabulary for Indexing and Retrieving the Literature of Water Resources Research and Development*. 3d ed. Washington, DC: U.S. Government Printing Office, 1980 (I1.2:W29/2/980) (PB 81-198376).

U.S. National Library of Medicine. *Medical Subject Headings: Annotated Alphabetic List, 1982* (July 1981), 822p. (PB 81-227480). Also: *Medical Subject Headings: Tree Structures, 1982* (PB

81-227472) and *Permuted Medical Subject Headings, 1982* (PB 81-227464).

TREATISES

Treatises present comprehensive information on the topics they cover and are oriented to researchers and advanced students. A treatise evaluates what is known and determines what facts are established in the subject area and points out what gaps of knowledge exist. Scientists often rely heavily on the treatises since in many cases they obviate the need for searching the older literature. The following are examples of treatises:

Alfvén, Hannes, and Gustaf Arrhenius. *Evolution of the Solar System.* Washington, DC: U.S. Government Printing Office, 1976 (NAS1.21:234).

Coatney, G. Robert, et al. *The Primate Malarias.* Washington, DC: U.S. Government Printing Office, 1971 (HE20.3252:M29).

REFERENCES

1. Krishna Subramanyam, *Scientific and Technical Information Resources* (New York: Marcel Dekker, 1981), pp. 7-8.

2. Rao Aluri, "Reference Sources among NTIS Technical Reports," *Reference Services Review* 5 (July/September 1977): 27-32; 6 (April/June 1978): 53-56; and Rao Aluri, "NTIS Reference Sources," *RQ* 19 (Summer 1980): 331-37.

INDEX